Elzbieta Maria Bitner-Gregersen
Lars Ingolf Eide · Torfinn Hørte
Rolf Skjong

Ship and Offshore Structure Design in Climate Change Perspective

 Springer

Elzbieta Maria Bitner-Gregersen
Lars Ingolf Eide
Torfinn Hørte
Rolf Skjong
Det Norske Veritas
Høvik
Norway

ISSN 2213-784X   ISSN 2213-7858   (electronic)
ISBN 978-3-642-34137-3   ISBN 978-3-642-34138-0   (eBook)
DOI 10.1007/978-3-642-34138-0
Springer Heidelberg New York Dordrecht London

Library of Congress Control Number: 2012956377

Printed on acid-free paper

Springer is part of Springer Science+Business Media (www.springer.com)

Foreword

Safety at sea is one of the main concerns of shipping and offshore industries in general and Classification Societies in particular. New designs must be assessed and operational decisions/made relative to recognized codes and standards. This will be the responsibility of the relevant authority, the classification society or the user himself, depending on the design and its application.

The importance of including the state-of-the-art knowledge about meteorological and oceanographic (met-ocean) conditions in ship and offshore standards has been in focus in the ship and offshore industry for many years. It has been recognized that there are potential safety, economic, and environmental advantages in utilizing the recent knowledge about met-ocean description in standards' development. To achieve recognition, a met-ocean description must be demonstrated to be robust and of adequate accuracy. Updating codes and standards takes, as most formal processes, some time, and consequently, the updates may lag a little behind the state-of-the-art.

The observed and projected climate changes in the past decades have been followed closely by Det Norske Veritas AS (DNV) and Classification Societies in general. The International Association of Classification Societies (IACS) has met-ocean conditions and variability on the agenda, and DNV is supporting IACS by several activities on adaptation process to climate change. There are still significant uncertainties related to climate change predictions and revisions of rules and standards may seem premature. However, we are concerned with knowing what impact climate changes in met-oceanographic conditions may have on future ship design and operations. Results reported in this monograph are a part of the investigations striving to shed light on the topic.

I endorse the work of the authors and have the pleasure to recommend this monograph for the reader.

Høvik, August 2012
Tor Svensen
DNV President

SpringerBriefs in Climate Studies

For further volumes:
http://www.springer.com/series/11581

Preface

Global warming and extreme weather events reported in the past years have attracted a lot of attention in academia, industry, and media. The ongoing debate around the observed climate change has focused on three important questions: will occurrence of extreme weather events increase in the future, which geographical locations will be most affected, and to what degree will climate change have impact on future ship traffic and design of ships and offshore structures?

The present study shortly reviews the findings of the Intergovernmental Panel on Climate Change Fourth Assessment Report, AR4, (IPCC 2007), the IPCC SREX "Summary for Policymakers" report (IPCC 2012) and other relevant publications regarding projections of meteorological and oceanographic conditions in the twenty-first century and beyond with design needs in focus. Emphasis is on wave climate and its potential implications on safe design and operations of ship structures.

The reviewed studies agree that there has been an increase in significant wave height (SWH) from the middle of the twentieth century to the early twenty-first century in the northern hemisphere winter in high latitudes in the north Atlantic and the north Pacific, with a decrease in more southerly latitudes of the northern hemisphere. The increase of the 99-percentile SWH has been observed to be 0.5–1.0 % per year (Young et al. 2011). However, if the record is extended back to late nineteenth century the picture changes, as studies show that storminess and wave heights in late nineteenth/early twentieth century were about the same as near the end of the twentieth century (Gulev and Grigorieva 2004). Thus, it is unclear if the increase observed during the past 4–5 decades is caused by anthropogenic climate change or just manifestation of long-term natural variability.

It is uncertain how future climate change will impact the extreme sea states that will be encountered by ocean going vessels. The reviewed studies show that there will be regional increases in the wind speeds and wave heights, more pronounced for the extremes than for the means. The increases of the 20-year return period of SWH or the highest storms in 20–30 years intervals are generally in the range 0.5–1.0 m in the North and Norwegian Seas, immediately west of the British Isles, off the northwest of Africa, around 30°N from the east coast of the United states to 50°W and in the Pacific between 25 and 40°N and from the west coast of the

United States to 170°W. However, increases up 18 % for the 99th percentile SWH have been reported for the southern North Sea by one paper of Grabemann and Weisse (2008). Thus, the increase may reach more than 10 % above present day extremes in some areas. The projections are influenced by choice of climate model, emission scenario, and downscaling method for waves. The uncertainty of the estimated increases is of the same order as the estimates.

To account for climate change of met-ocean conditions and uncertainty connected to the estimates of future extreme wave heights and other met-ocean parameters a risk-based approach for ship and offshore structure design is proposed. The impact of expected wave climate change on ship design is demonstrated for five differently sized oil tankers, ranging from Product Tanker of length 175 m to VLCC with length 320 m. The presented examples show consequences of climate change for the hull girder failure probability. They demonstrate that in order to maintain the safety level the steel weight of the deck in the midship region should be increased by 5–8 % if the extreme SWH increases by 1 m. However, it should also be noted that weather forecasts are improving, and ships ability to avoid extreme met-ocean conditions by using weather routing systems may imply that the current practice of designing ships to 20-years' North Atlantic extreme conditions may be relaxed in the future.

Recommendations for future research activities allowing the marine industry to adapt to climate change are given.

References

Grabemann I, Weisse R (2008) Climate change impact on extreme wave conditions in the North Sea: an ensemble study. Ocean Dyn 58:199–212

Gulev SK, Grigorieva V (2004) Last century changes in ocean wind wave height from global visual wave data. Geogr Res Lett 31:L24302. doi:10.1029/2004GL021040

IPCC (2007) Climate change. The physical science basis. In: Solomon S, Qin D, Manning M, Chen Z, Marquis M,Averyt KB, Tignor M, Miller HL (eds) Contribution of working group I to the fourth assessment report of the intergovernmental panel on climate change. Cambridge University Press, Cambridge, p 996

IPCC (2012) Managing the risks of extreme events and disasters to advance climate change adaptation. In: Field CB, Barros V, Stocker TF, Qin D, Dokken DJ,Ebi KL, Mastrandrea MD, Mach KJ, Plattner G-K, Allen SK, Tignor M, Midgley PM (eds) A special report of working groups I and II of the Intergovernmental Panel on climate change. Cambridge University Press, Cambridge, p 582

Young RI, Zieger S, Babanin AV (2011) Global trends in wind speed and wave height. Science 332(22):451–455

Acknowledgments

This work, initiated in 2008, was carried out within the internal DNV Research and Innovation research activities in the DNVR&I Maritime Transport Programme. Our thanks go to the Programme Director Dr. Øyvind Endresen for his encouragement during the writing of this monograph. The application of the risk-based approach using structural reliability analysis builds on research results from the strategic research programmes over a period of two decades, and makes the integration of new information on met-ocean conditions in the Rule development easy and consistent. The continuous support is highly appreciated. We would also like to thank the Norwegian Meteorological Institute for providing the NORA10 hindcast and ERA-Interim data, the anonymous reviewers for valuable comments and several publishers for permissions to reproduce pictures and images. The opinions expressed herein are those of the authors and should not be construed as reflecting the views of the company.

August 2012, Høvik

Elzbieta Maria Bitner-Gregersen
Lars Ingolf Eide
Torfinn Hørte
Rolf Skjong

Acknowledgements

Contents

Chapter 1
Introduction

Safety at sea is one of the main concerns of shipping and offshore industry in general and Classification Societies as well as oil companies in particular. The importance of including the state-of-the-art knowledge about meteorological (temperature, pressure, wind) and oceanographic (waves, current) conditions in ship standards have been discussed increasingly by industry and academia in the last decades in several international forums. There are potential safety, economic, and environmental advantages in utilizing the most recent knowledge about meteorological and oceanographic (met-ocean) conditions and investigating its implication for design and operation of ship and offshore structures.

The ongoing debate around the observed and projected climate change has confronted the shipping and offshore industry with two important questions: Is it likely that ship and offshore structures will experience higher environmental loads; and Will Classification Societies' Rules and Offshore Standards need to be updated? The present study makes an attempt to answer these questions based on the state-of the-art knowledge about climate change and structural reliability analysis.

In this monograph emphasis is on wave climate, which is expected to have the largest impact on ship and offshore structure design in comparison to other environmental phenomena. Changes in wind climate may affect also loads and responses of ship and offshore structures, depending on how significant they will be, while projected changes in sea level combined with potential increases in storm surge activity have little potential to influence ship design directly but are expected to have impact on harbours, fixed offshore structures and coastal installations, e.g. on harbour depths and offloading and deck heights. Secondary effects, such as a possible increase in marine growth due to warmer oceans may increase loads on ship and offshore structures in some ocean regions, e.g. the Baltic Sea. However, this effect may also be compensated by improved antifouling coating.

We start with a short review of the findings of the Intergovernmental Panel on Climate Change Fourth Assessment Report, AR4, (2007), the IPCC Special Report on Managing the Risks of Extreme Events and Disasters to Advance Climate Change Adaptation (SREX) (IPCC 2011, 2012) and other publications regarding

E. M. Bitner-Gregersen et al., *Ship and Offshore Structure Design in Climate Change Perspective*, SpringerBriefs in Climate Studies, DOI: 10.1007/978-3-642-34138-0_1, The Author(s) 2013

projections of met-ocean conditions in the twenty-first century and beyond. We also illustrate the impact relevant uncertainties may have on climate change projections with design needs in focus.

It is emphasized that this review of expected impacts of anthropogenic climate change on the wind and wave conditions in the twenty-first century is limited to look for evidence in the scientific literature. It is not a critical scientific review of the publications with respect to methods, use of data or similar, neither is it discussing whether anthropogenic climate changes are happening or likely to happen in the future, but just asking: if it happens, what changes in wind and wave conditions can be expected according to recent published information and what are the possible impacts for ship transport and offshore structures? It is not an exhaustive review of all possible publications that deal with the topic, just a selection of key references.

Although the presented review is not covering all studies regarding climate change the authors believe that the reader will gain a fair and balanced view of the state-of-the-art in the field of climate change and would be able to understand the importance of the existing findings for design and operations of ships and also offshore structures in general.

Another limitation of this monograph is that it has not covered all combinations of extreme weather types and structures, as illustrated in the matrix below. Our focus has been on ships and extra-tropical cyclones.

Weather type	Structure	
	Ship	Offshore platforms
Extra-tropical cyclones ("Regular storm")	Increase in wave height by region, examples of impact	Increase in wave height by region
Tropical cyclones (hurricanes and typhoons)	General statements on potential changes in intensity and frequency, no regional information. Tropical cyclones may generally be avoided by ships, and offshore operations may be closed down	

We show how the latest scientific results on climate change can be in-cooperated in design practice of ship and offshore structures. A risk based approach that continuously allows combining new information about climate change and relevant uncertainties in ship and offshore structure design is proposed being an extension of a systematic approach built up over many years. Further, the potential impact of wave climate change on ship structure design is demonstrated for five oil tankers, ranging from Product Tanker to Very Large Crude Oil Carrier (VLCC). Consequences of climate change for the hull girder failure probability and hence the steel weights (reflecting potential increased of costs) needed to compensate for the increase of the failure probability in the midship deck region are shown. Recommendations for future research activities which will allow adaptation of the shipping and offshore industry to climate change are given.

References

IPCC (2007) Climate change. The Physical science basis. In: Solomon SD, Qin M, Manning Z, Chen M, Marquis KB, Averyt M, Tignor and Miller HL (eds) Contribution of working group I to the fourth assessment report of the intergovernmental panel on climate change. Cambridge University Press, Cambridge,. pp 996

IPCC (2011) Summary for policymakers. In: Field CB, Barros V, Stocker TF, Qin D, Dokken D, Ebi KL, Mastrandrea MD, Mach KJ, Plattner G-K, Allen S, Tignor M, Midgley PM (eds) Intergovernmental panel on climate change special report on managing the risks of extreme events and disasters to advance climate change adaptation. Cambridge University Press, Cambridge

IPCC (2012) Managing the risks of extreme events and disasters to advance climate change adaptation. In: Field CB, Barros V, Stocker TF, Qin D, Dokken DJ, Ebi KL, Mastrandrea MD, Mach KJ, Plattner G-K, Allen SK, Tignor M, Midgley PM (eds) A special report of working groups I and II of the intergovernmental panel on climate change. Cambridge University Press, Cambridge, pp 582

Chapter 2
Observed and Predicted Climate Change

2.1 IPPC Scenarios

The Intergovernmental Panel on Climate Change (IPCC) was established jointly by the World Meteorological Organization (WMO) and the United Nations Environment Programme (UNEP) in 1988. The mandate was to assess scientific information related to climate change, to evaluate the environmental and socio-economic consequences of climate change, and to formulate realistic response strategies. The assessments provided by IPCC have since then played a major role in assisting governments to adopt and implement policies in response to climate change. In particular the IPCC has responded to the need for authoritative advice of the Conference of the Parties (COP) to the United Nations Framework Convention on Climate Change (UNFCCC), which was established in 1992, and its 1997 Kyoto Protocol. Since its establishment in 1988, the IPCC has produced a series of Assessment Reports (1990, 1995, 2001 and 2007). All Assessment Reports consist of three parts: The Science of Climate Change, Impacts, Adaptations and Mitigation of Climate Change: Scientific-Technical Analyses and Economic and Social Dimensions of Climate Change. The last two also include a Synthesis report. In addition, IPCC Special Reports, Technical Papers and Methodology Reports have become standard works of reference, widely used by policymakers, scientists, other experts and students, e.g. the "Special Report on Managing the Risks of Extreme Events and Disasters to Advance Climate Change Adaptation" (IPCC 2012, hereafter called SREX. A "Summary for Policy Makers" was issued by IPCC in 2011).

The IPCC 2007 Fourth Assessment Report on Climate Change (hereafter called AR4; IPCC 2007a) provides information for policymakers, scientists and engineers on the current understanding of scientific, technical and socio–economic aspects of climate change. It consists of the following sub-reports:

E. M. Bitner-Gregersen et al., *Ship and Offshore Structure Design in Climate Change Perspective*, SpringerBriefs in Climate Studies, DOI: 10.1007/978-3-642-34138-0_2, The Author(s) 2013

- The AR4 Synthesis Report.
- The Working Group I Report "The Physical Science Basis", hereafter AR4.1.
- The Working Group II Report "Impacts, Adaptation and Vulnerability".
- The Working Group III Report "Mitigation of Climate Change".

A summary of the findings of the three Working Group reports can be found in the AR4 Synthesis Report (IPCC 2007a), which specifically addresses the issues of concern to policymakers in the domain of climate change. It confirms, with little uncertainty, that climate changes observed now are mostly a result of human activities. The report illustrates the impacts of global warming being already under way and to be expected in the future, and describes the potential for adaptation of society to reduce its vulnerability. Finally, it presents an analysis of costs, policies and technologies intended to limit the extent of future changes in the climate system.

In the following we will only consider the results from the Working Group I Report "The Physical Science Basis" (IPCC 2007b).

The extent to which human activities will impact future climate conditions depend to a high degree on how the international society reacts to the prospect of significant global warming with its consequences for changes in regional and local climate. AR4 considered four scenario classes based on various socio–economic developments and their impacts on emissions of greenhouse gases. Special attention was given to issues of human well-being and development. Technologies, policies, measures and instruments as well as barriers to implementation were addressed in the AR4 reports along with synergies and trade-offs.

The scenarios adopted by AR4 have been used to project climate changes in the 21st century and beyond. The A* scenarios are pessimistic ones (higher increase of the Earth surface temperature) while the B* scenarios are optimistic ones with respect to reduction of greenhouse gases. Thus choice of a scenario will affect results and introduce uncertainties in climate change projections.

A1. This family describes a future world of very rapid economic growth and a global population that peaks in mid-century and declines thereafter. It also assumes rapid introduction of new and more efficient technologies. The A1 family has three sub-groups that describe alternative directions of technological change in the energy system: A fossil-intensive one (A1FI); one based on non-fossil energy sources (A1T); and one that is a balance across all sources (A1B).

A2. This family describes a very heterogeneous world. Selfreliance and preservation of local identities are important factors and the population increases continuously. Economic development is primarily regionally oriented and per capita economic growth and technological change more fragmented and slower than other storylines.

B1. This family describes a convergent world with a global population, that peaks in mid-century and declines thereafter, as in the A1 family, but with economic structures that change rapidly toward a service and information economy. The intensity in material consumption is reduced and clean and resource-efficient technologies are introduced. The emphasis is on global solutions to economic,

Fig. 2.1 Projected changes
of global surface warming.
Lines are means, *shading*
are + one standard deviation
of individual model annual
means. From IPCC (2007b)

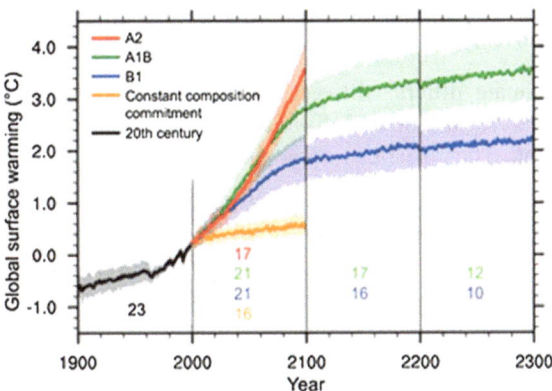

social and environmental sustainability, including improved equity, but without
additional climate initiatives.

B2. In the B2 family the global population is continuously increasing but at a
rate lower than A2. Emphasis is on local solutions to economic, social and
environmental sustainability, intermediate levels of economic development, and
less rapid and more diverse technological change than in the B1 and A1 storylines.
While the B2 family of scenarios is oriented towards environmental protection and
social equity, it focuses on local and regional levels.

In addition to these four scenario families some of the reviewed papers also
include the IPCC IS92a scenario (IPCC 2001), in which the concentration of CO_2
in the atmosphere effectively increases by 1 %/year from 1990. This is often
considered to be the "business as usual" scenario.

Figure 2.1 shows the multi-model means of global surface warming relative to
1980–1990 for three of the scenarios (IPCC 2007b).

Table 2.1 shows the projected global average surface warming and range for
the scenarios mentioned above (IPCC 2001, 2007b).

Table 2.1 Projected global
average surface warming
and range for the scenarios
referred later in the text.
(Based on IPCC 2001
and 2007b)

Scenario	Projected temperature change (°C increase between 1980–1999 and 2090–2099)	
	Best estimate	Likely range
Constant CO_2-concentration at 2000 level	0.6	0.3–0.9
IS92a (IPCC 2001)	2.3	1.9–3.4
B1	1.8	1.1–2.9
A1T	2.4	1.4–3.8
B2	2.4	1.4–3.8
A1B	2.8	1.7–4.4

2.2 Climate Change and Variability

Climate differs with geographic location and is influenced by, amongst other factors, latitude and distance from the oceans. Climate has always changed with time. The variations observed today are due to:

- Natural variability, originating from the internal dynamics of the Earth's system and occurring usually on time scales a few years via decadal to multi-decadal, but much longer cycles due to movement of poles may also occur, e.g. 21000 year Milankovitch cycles.
- Climate change due to external forcing, such as changes in solar radiation and volcanic activity, varying on time scales from years to millennia.
- Anthropogenic climate change, caused by human activities and in particular emissions of greenhouse gases (GHG), which takes place over a few decades to centuries.

In the AR4.1 report the IPCC analysed the chain from GHG emissions and concentrations, via radiative forcing and to potential resultant climate change. The set of AR4 reports also evaluated to what extent observed changes in climate and in physical and biological systems can be attributed to natural or anthropogenic causes. It was concluded that warming of the climate system is unequivocal, as it is now evident from observations of increases in global average air and ocean temperatures, widespread melting of snow and ice and rising global average sea level. According to the AR4.1 report there is very high confidence that the net effect of human activities since 1750 has contributed significantly to the global warming. Global GHG emissions due to human activities have grown since pre-industrial times, with an increase of 70 % between 1970 and 2004. The global GHGs emission needs to be reduced significantly before 2030, in order to limit the warming to 2 °C (IPCC 2007a).

The SREX report (IPCC 2012) assessed the scientific literature, including in particular investigations carried out after the AR4.1 report was issued. Emphasis is on the relationships between climate change and extremes of weather events and the implications for society. The SREX operates with degrees of confidence in observed trends and projections, i.e. low, medium and high, as well as with various degrees of likelihood that there will be certain developments, ranging from "exceptionally unlikely" to "virtually certain". It is pointed out that assigning "low confidence" in a specific extreme on regional or global scale neither implies nor excludes the possibility of changes in this extreme. Many uncertainties remain in modern climate change projections.

In the last decades increasing attention has been given to climate change induced by human activities, its interaction with natural climate variability, and possible consequences for design. It is, however, important to be aware that the natural climate variability can be of the same order of magnitude as the anthropogenic climate change and may mask it for several years to come.

Extreme wind speeds and other attributes of wind represent potential threats to human safety and human activities on land, at sea and in the air. Winds are the driver of ocean waves and trends in average wind speeds may result in feedback on the climate, e.g. in terms of increased evaporation. Unfortunately, AR4.1 does not go into much detail regarding wind and wave conditions and addresses these topics mainly in terms of tropical and extra-tropical cyclone activity. However, in SREX these topics are addressed in more detail. It is reported, for the first time, how expertise in climate science, disaster risk management, and adaptation can work together to inform discussions on ways to reduce and manage the risks of extreme events in a changing climate. The report assesses the impact climate change have had and may continue to have in altering characteristics of extreme events, as well as experience gained by institutions, organizations, and communities to mitigate impacts of climate extremes. Among these are early-warning systems, improvements in infrastructure, and the expansion of social safety nets.

The SREX report provides information on how natural climate variability and human-generated climate change influence the frequency, intensity, spatial extent, and duration of some extreme weather and climate events. Case studies that illustrate specific extreme events and their impacts in different parts of the world are also included in SREX as well as a range of risk management activities.

Below we provide in more detail results from a selection of what we consider to be key publications referred to in SREX that address observed changes in wind speed and wave heights since the late 1880'ies until around 2005 (Sects. 2.3 and 2.4) and the changes predicted for the 21st century (Sects. 2.5 and 2.6). The publications, which have been selected in the context of climate change impact on shipping, are supplemented by publications issued after SREX as well as by other publications addressing changes in wind and wave climate.

It must be noted that the majority of the papers reviewed herein have been academic/scientific papers not written with the needs of the designer in mind. Therefore the extreme values presented there are not necessarily directly applicable to engineering design practices. To proper assess the impact of climate change on wind and wave conditions, with estimates of changes in design values and the associated uncertainties, the shipping and offshore industry would need access to the raw data in terms of time series.

2.3 Changes in Storminess and Wind in the Twentieth Century

2.3.1 Extra-Tropical Storms

When trying to establish trends of mean and extreme wind conditions one should be aware of a few factors:

1. As pointed out by e.g. IPCC (2012) long-term high-quality wind measurements from terrestrial anemometers are sparse in many parts of the globe and the measurements are influenced by changes in instrumentation, station location, and surrounding land. This has hampered the direct investigation of changes in wind climatology.
2. The observations of marine winds have been hampered by inadequate instrumentation and inhomogeneous records. The longest records are surface wind and meteorological observations from Voluntary Observing Ships (VOS), which became systematic around 150 years ago and are assembled in ICOADS (Worley et al. 2005). Apparent significant trends in scalar wind should be considered with caution as VOS wind observations are influenced by time-dependent biases, resulting from the rising proportion of anemometer measurements, increasing anemometer heights, changes in definitions of Beaufort wind estimates, growing ship size, as well as inappropriate evaluation of the true wind speed from the relative wind and time-dependent sampling biases (see e.g. IPCC 2007b for references).
3. Reliable global wind data from satellites go back only a few decades, to the mid-1980'ies
4. An important source of decadal changes in storminess and wind speed is reanalysis of weather maps. Such information can be used back to the 1950'ies.
5. There may have been local and regional differences in changes and trends.
6. Natural variability occurs on several time scales and it will generally not be sufficient to consider only the last 30–50 years.
7. To go back before 1950 will generally require proxy data, e.g. geostrophic winds calculated from pressure data.

Despite a noticeable increase in global surface temperature the last 50–60 years AR4.1 did not identify any significant global trends in average marine wind speeds. It appears, however, that there are regional differences and that wind speed has shown an upward trend in the tropical North Atlantic and extra-tropical North Pacific and downward trends in the equatorial Atlantic, tropical South Atlantic and subtropical North Pacific (AR4.1, Sect. 3.5.6). This was based on considerations of time series of local surface pressure gradients. AR4.1 also reported that changes in the large-scale atmospheric circulation are apparent and that mid-latitude westerly winds have generally increased in both hemispheres. Furthermore, AR4.1 described evidence for a poleward shift in storm tracks, with resulting increase in wind speeds in the North Pacific and the North Atlantic. This latter finding has been corroborated in SREX, which cites studies published between 2007 and 2010.

The AR4.1 did not specifically address changes in extreme wind although it did report on wind changes in the context of other phenomena such as tropical and extra-tropical cyclones and oceanic waves. The topic of trends in wind speed and storminess, and particularly extreme wind speeds, are dealt with in more detail in SREX (IPCC 2012), which refers to a large number of publications that have considered changes and trends in wind conditions. The majority of the references deals with changes over land on regional scales, and there seems to be more publications that indicate declining wind speeds than vice versa.

The number of studies that consider trends in storm activity or trends in wind speed on a global scale over longer periods than four to five decades is rather limited. The North and the Northeast Atlantic appear to be the regions that have been most extensively studied with respect to historic storm activity. These are also regions that have frequently been used as reference for ship design. Reliable pressure data exist for several stations along the coasts of the North Sea, the Norwegian seas, the Faeroe Isles, Iceland, Ireland, Jan Mayen, mainland Northern Europe, Greenland and the weather ship "M". Several of these data sets have been used to create proxies for wind by using pressure tendencies and geostrophic winds calculated from triangles of pressure observations. The Wasa Group (1998) was set up to verify or disprove the hypotheses that storm and wave climate in the northeast Atlantic and its adjacent seas have worsened in the present century. They used pressure data from a range of the mentioned stations to calculate geostrophic winds between 1881 and 1995. The study showed that storm and wave climate in most of the northeast Atlantic and in the North Sea did indeed roughen in last decades of the 20th century. However, a significant conclusion is that the storm and wave climate has undergone significant variations on timescales of decades, and that the present intensity of the storm and wave climate seems to be comparable with that at the end of the 19th century and beginning of the 20th. Part of this variability is found to be related to the North Atlantic oscillation. The study was extended by three years, to 1998, by Alexandersson et al. (2000).

Wang et al. (2009a) extended the data even further to 1874–2007 and used slightly different data preparation methods than Alexandersson et al. (2000). However, they confirmed the results, and these are illustrated in Fig. 2.2.

The studies referenced above suggest that

- There was relatively high storminess around 1900 and in the 1990s.
- The 1960s and 1970s were periods of low storm activity.

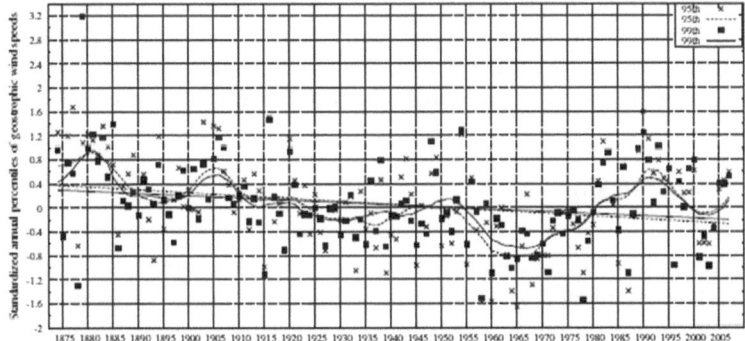

Fig. 2.2 NE Atlantic region area averages of standardized annual 99th and 95th percentiles of 3-hourly geostrophic wind speeds, and the corresponding Gaussian low-pass filtered curves and linear trends. (From Wang et al. 2009a)

- Storminess in the NE Atlantic exhibits strong inter-decadal variability.
- The latter half of the 20th century was punctuated by a peak in storminess around 1990 which according to Wang et al. (2009a) is unprecedented since 1874.

Bärring and von Storch (2004) and Bärring and Fortuniak (2009) went even further back, to 1780, without finding any long-term trends in storminess.

One study, Young et al. (2011), looked at global marine winds for the period 1991–2008 using altimeter from seven different missions of GEOSAT. Although GEOSAT was launched in 1985 only data from 1991 were used for the wind analysis due to questionable data quality for the first six years. Wind mean speed as well as the 90th and 99th percentiles were calculated. The results are shown in Fig. 2.3.

Young et al. (2011) found that with a few exceptions all three wind parameters (mean, 90 and 99 %percentiles) increased over the world oceans from 1991 to 2008. The study reports that the mean and 90th percentile wind have increased by at least 0.25–0.5 % per year with stronger trend in the Southern than in the Northern Hemisphere, apart from the central north Pacific. The 99th percentile extreme wind speeds have increased over the majority of the world oceans by at least 0.75 % per year. The trends are not statistically significant everywhere (point that are statistically significant are marked with dots in Fig. 2.3). Young et al. (2011) compared the trends from the GEOSAT data to in situ measurements from buoys and numerical model results and found qualitatively consistent results. They point out that the validity of the study is limited to the period 1991–2008 and that the observed trends are not necessarily the result of global warming.

Young et al. (2011) found that with a few exceptions all three wind parameters (mean, 90 and 99 %percentiles) increased over the world oceans from 1991 to 2008. The study reports that the mean and 90th percentile wind have increased by at least 0.25 to 0.5 % per year with stronger trend in the Southern than in the Northern Hemisphere, apart from the central north Pacific. The 99th percentile extreme wind speeds have increased over the majority of the world oceans by at least 0.75 % per year. The trends are not statistically significant everywhere (point that are statistically significant are marked with dots in Fig. 2.3). Young et al. (2011) compared the trends from the GEOSAT data to in situ measurements from buoys and numerical model results and found qualitatively consistent results. They point out that the validity of the study is limited to the period 1991–2008 and that the observed trends are not necessarily the result of global warming.

2.3.2 Tropical Storms

AR4.1 (IPCC 2007b) argued that is more likely than not that anthropogenic influence has contributed to an increase of the most intense tropical cyclones, a common name for hurricanes and typhoons. However, this view was challenged by the Sixth WMO International Workshop on Tropical Cyclones (IWTC-VI), which in its proceedings (WMO 2007) included the following points amongst its consensus statements:

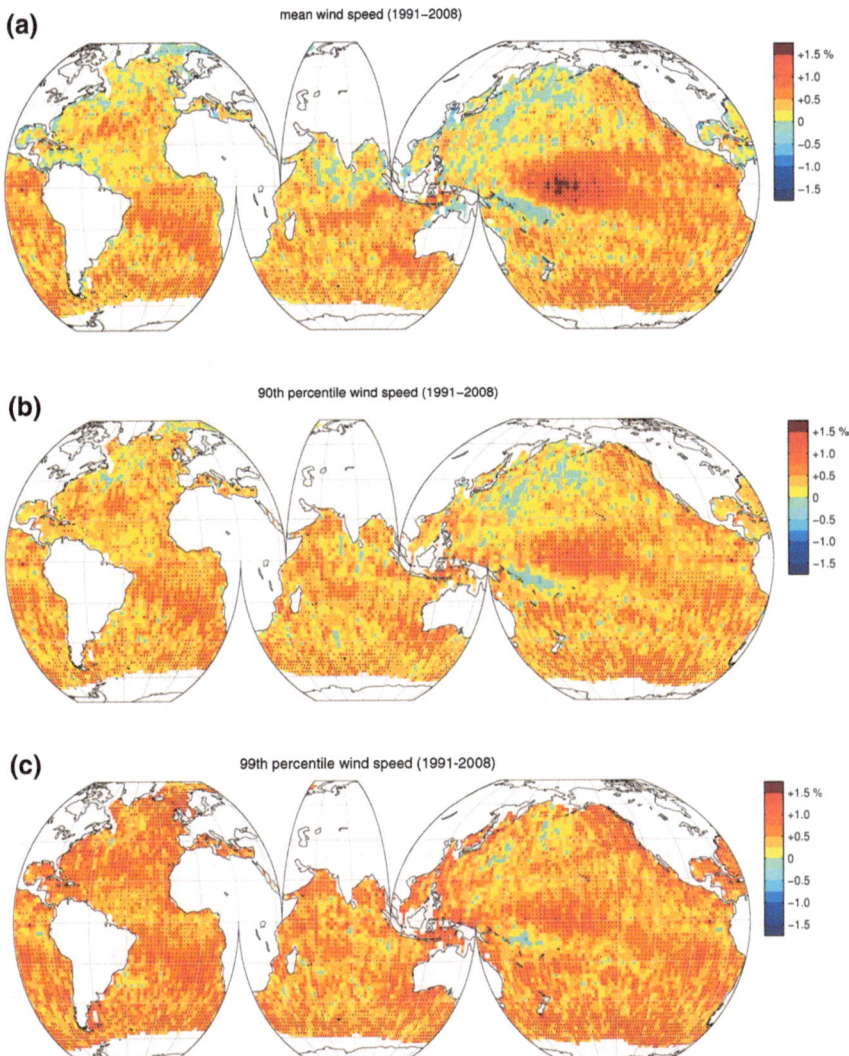

Fig. 2.3 Changes in **a** mean wind speed **b** the 90th and **c** the 99th for the period 1991–2008 as found from GEOSAT altimeter. *Dots* indicate statistically significant results (From Young et al. 2011; published with permission from American Association for the Advancement of Science, AAAS)

1. Though there is evidence both for and against the existence of a detectable anthropogenic signal in the tropical cyclone climate record to date, no firm conclusion can be made on this point.
2. No individual tropical cyclone can be directly attributed to climate change.
3. There is an inconsistency between the small changes in wind-speed projected by theory and modelling versus large changes reported by some observational studies.

4. The recent increase in societal impact from tropical cyclones has largely been caused by rising concentrations of population and infrastructure in coastal regions.
5. Tropical cyclone wind-speed monitoring has changed dramatically over the last few decades, leading to difficulties in determining accurate trends.
6. Large regional variations exist in methods used to monitor tropical cyclones. Also, most regions have no measurements by instrumented aircraft. These significant limitations will continue to make detection of trends difficult.
7. There is an observed multi-decadal variability of tropical cyclones in some regions whose causes, whether natural, anthropogenic or a combination, are currently being debated. This variability makes detecting any long-term trends in tropical cyclone activity difficult.

This conclusion has gained increased confidence through the work of e.g. Knutson et al. (2010), who summarized progress since 2006 by stating that "Therefore, it remains uncertain whether past changes in tropical cyclone activity have exceeded the variability expected from natural causes". Knutson et al. (2010) argued that substantial limitations in the availability and quality of global historical records of tropical cyclones as well as large amplitude fluctuations in the frequency and intensity of tropical cyclones greatly impede and complicate the detection of long-term trends and their attribution to rising levels of atmospheric greenhouse gases.

For a continuous update of hurricane and climate change, see e.g.the hurricane portal of Geophysical Fluid Dynamics Laboratory (GFDL): http://www.gfdl.noaa.gov/hurricane -portal.

2.4 Changes in Waves in the Twentieth Century

It is not necessarily the wind speed as such that is of most interest to seafarers or the coastal populations but rather the waves and storm surges the wind generates. Here the focus is on waves and in particular wave heights. Other parameters like wave spectra which show the distribution of wave energy with direction and frequencies or periods are seldom addressed in publications on the effects of a changing climate on the wave conditions. The most common metrics are means, certain percentiles or number of events above certain levels of significant wave height, SWH (H_s). SWH is traditionally defined as the average height from trough to crest of the highest one-third of waves in a sea state. With the introduction of analogue and digital wave recording devices the most commonly used definition now is four times the standard deviation of the surface elevation or equivalently as four times the square root of the first moment of the wave spectrum, valid for the Gaussian sea surface (Longuet-Higgins 1952; Cartwright and Longuet-Higgins 1956). The difference in magnitude between the two definition will usually be limited to only a few percent, see e.g. Bitner-Gregersen and Hagen (1990). It is important to be aware that the highest wave in a sea state will be larger than the

SWH, in theory one might encounter a wave that is up to double the significant wave height and even higher if rogue waves are present.

The definition of a sea state is not absolute. General requirements are that the wave conditions should remain stationary during a sea state and that the duration has to be much longer than the individual wave period, but smaller than the period in which the wind and swell conditions vary significantly. Typically, records of a few hundred to a few thousand wave periods are used to determine the wave statistics for a sea state. Note that in design 3 or 6 h duration is commonly adopted. An assumed sea state duration will influence extreme statistics of individual wave parameters. Probability of extreme individual wave events will increase with increase sea state duration (Longuet-Higgins 1952).

Gulev and Grigorieva (2004) pointed to increasing annual mean significant wave heights in the North Pacific and North Atlantic by 0.05–0.1 m/decade and negative trends in the tropical western Pacific, south Indian Ocean and the Tasman Sea from 1950 to 2001. However, if the records are extended back to the late 19th century the picture changes somewhat. For the northeast Atlantic (44°N–52°N, 6°E–20°E) Gulev and Grigorieva (2004) found no trend between 1885 and 2002; in fact the highest annual mean significant waves as observed from ships were 0.1–0.15 m higher around 1925 and 1945 than in the 1990'ies. For the northeast Pacific (48°N–52°N, 132°W–146°W) the upward trend for 1885–2002, while still statistically significant, became considerably weaker than for the period 1950–2002 and the highest annual means for the first half of the period 1885–2002 are comparable to those of the last five decades. For the Northeast Atlantic these results are consistent with results quoted in Wang et al. (2009a) and those found by the Wasa Group (1998).

For the winter season Wang and Swail (2006a) found an upward trend in mean winter significant wave heights in the North Pacific between the 1958 and 1997 of 0.004–0.18 m/decade. The picture for the North Atlantic was more complicated. There the annual mean significant wave height was found to have increased by 0.10–0.30 m/decade from west of the British Isles to the northern North Sea and to decline by up to 0.2 m/decade further south, between the coasts of the United States and North Africa. Similar trends have also been reported by others for the northeast Atlantic (e.g. Caires and Swail 2004; see also AR4.1).

Young et al. (2011) analysed changes in significant wave height from satellite altimetry for the years 1985–2008. They found nearly neutral development of mean significant wave heights during the period, with weak negative trends over large part of the northern hemisphere oceans (see Fig. 2.4a). Large regions of the north Pacific and north Atlantic show a weak negative trend (0.25 % per year), as do much of the equatorial regions of all oceanic basins. However, the southern hemisphere has a consistent weak positive trend of approximately 0.25 % per year. The buoy and wave model data supported these conclusions. With the exception of the locations marked with dots in Fig. 2.4a, the trends are not statistically significant.

More interesting than changes in mean significant wave heights are the changes in the higher portion of the sea states, as represented by e.g. the 99th percentile of

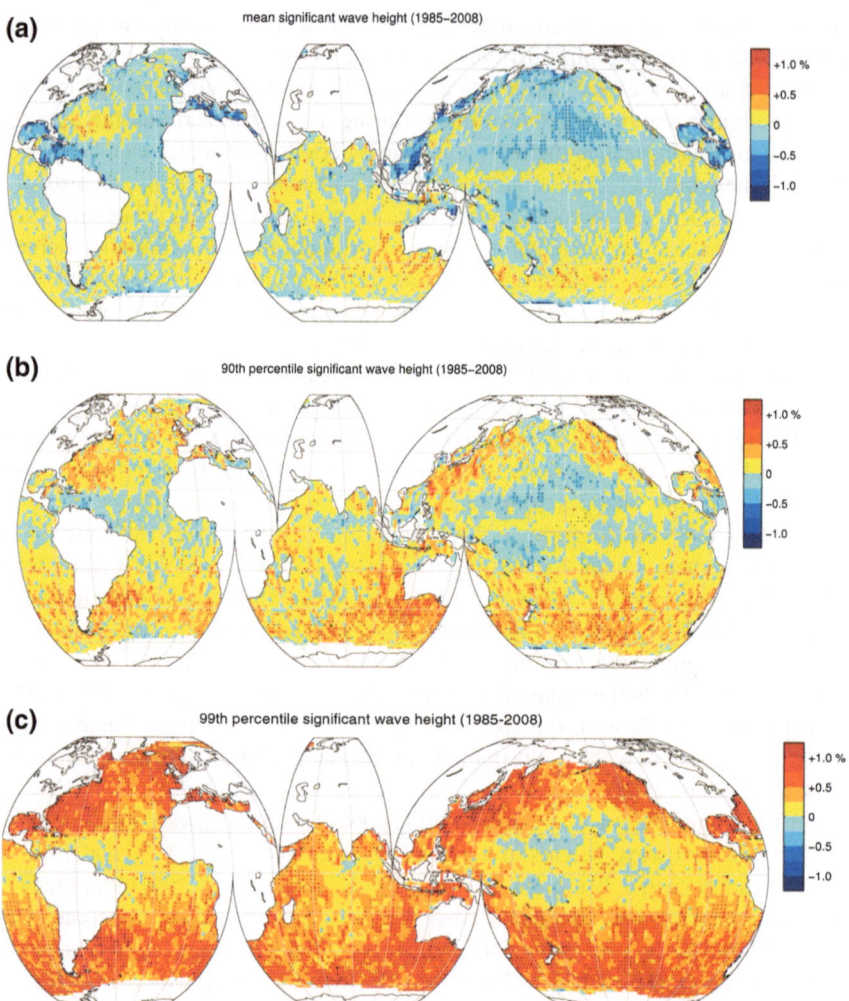

Fig. 2.4 Changes in **a** mean significant wave height **b** the 90th and **c** the 99th for the period 1985–2008 as found from GEOSAT altimeter. *Dots* indicate statistically significant results (From Young et al. 2011; published with permission from American Association for the Advancement of Science, AAAS)

the long-term distribution or the 20 year return period significant wave height. Several papers have addressed changes in these parameters. Wang and Swail (2006a) and Caires and Swail (2004) considered changes in extreme significant wave height since 1958. The former found that the pattern of change for the 20-year return period significant wave height had many similarities to the pattern of changes in the mean but was found to be more patchy. Furthermore, they found changes in the period 1958–1997 in the winter 20 year return period significant wave height of 0.2–1.2 m, almost three times as much as the increase in the winter mean significant wave height. The changes were found to be positive in the northeast Atantic and negative in the subtropical Atlantic. Also the Wasa Group (1998) reports more rapid growth in the annual maximum and 99th percentile significant wave heights than for the mean for the Northeast Atlantic.

From Fig. 2.4b and c it can be seen that Young et al. (2011) found that the 90th percentile and the 99th percentile wave height trends for the period 1985–2008 are progressively more positive than in the mean value, with the higher latitudes (above 35°) of both the hemispheres showing positive trends of approximately 0.25 % per year at the 90th percentile and 0.5–1.0 % at the 99th percentile.

Even though Pfizenmayer and Storch (2001) argued that a change in eastward propagating waves over the North Sea is a local manifestation of anthropogenic climate change, it is still very uncertain that such a signal is detectable on the extreme wave heights. Only one of the reviewed studies (Wang et al. 2009b) linked the observed changes in wave height during the last part of the 20th century to anthropogenic influence.

There is agreement between all the referenced studies that SWH has increased over the last half of the 20th century. However, it may be of interest to know if there have been differences in the changes between wind sea and swell over the last five decades. Gulev and Grigorieva (2006) used observations from Voluntary Observing Ships in the ICOADS database (Worley et al. 2005) to carry out a separate analysis for wind sea and swell. They found that in the northeast Atlantic these changes can be largely attributed to the variations in swell height, rather than to the wind sea. The same applied to some extent in the northeast Pacific. They also found that the inter-annual variability of wind sea and swell demonstrated noticeable differences in the North Atlantic and that they are more similar in the Pacific. Gulev and Grigorieva (2006) indicated that the SWH in the North Atlantic is more controlled by cyclone activity that directly by local wind speed. This is an important finding, as wind sea causes much of the danger for the operations of marine carriers, except for very large tankers and container ships, which can also suffer from the high swells even under calm conditions. Moreover, swell significantly affects the tankers approaching the oil platforms for bunkering.

2.5 Expected Changes in Storminess and Wind in the Twenty-first Century

2.5.1 Extra-Tropical Storms

Although the findings in AR4.1 were not conclusive the report found consistent support for continued poleward shift in storm tracks and greater storm activity at high latitudes in both hemispheres, with a reduction in the number of mid-latitude storms through the 21st century.

Post-AR4.1 studies support a poleward shift in tropospheric storm tracks. Bengtsson et al. (2009) used a single model approach to study the effects of a warming climate on storm activity, with emphasis on the northern hemisphere. They used a high-resolution version (63 km) of the Max Planck Institute Global Circulation Model (GCM) ECHAM5 to run two 32 years long periods at the end of the 20th and 21st centuries for the IPCC A1B emission scenario. Their general conclusion is that one can expect a small reduction in the number of cyclones without any significant changes in extreme winds in both hemispheres. Bengtsson et al. (2009) presented more detailed analysis for the northern hemisphere by looking at cyclones with wind speeds above 35 and 45 ms^{-1} at a height of 925 hPa and by looking at different regions. They found that a reduction in the number of cyclones with wind speeds above 35 ms^{-1} may be expected, except for the summer months June, July and August. However, the difference was hardly significant except for the Atlantic. There the model study showed a general reduction of cyclones >35 ms^{-1} over southern Europe, minor changes over northern Europe and an increase in the Arctic. They speculate that the increase in the Arctic in the 21st may be related to the reduced ice cover that provides more favourable conditions for high winds.

For storms with wind speed at 925 hPa > 45 ms^{-1} Bengtsson et al. (2009) found for all practical purposes no change in the number of storms summed over the northern hemisphere. However, there was one significant regional change. The number of these extreme storms increased over the British Isles and Scandinavia whereas it decreased near Greenland and towards the Mediterranean region. Another indication in Bengstsson et al. (2009), is that the total number of storms with wind speed >55 ms^{-1} is reduced both in the Atlantic and the Pacific, indicating that although the wind speed may increase somewhat in some regions the overall Atlantic extreme does not, as illustrated in Fig. 2.5 (from Bengtsson et al. 2009).

Finally, Bengtsson et al. (2009) noted an increase in the number storms with wind speed >45 ms^{-1} in the northern hemisphere during summer (June, July and August), particularly in the Pacific. They attribute this to storms with a purely tropical origin that migrate into lower and middle latitudes and undergo extra-tropical transition.

Bengtsson et al. (2009) and Champion et al. (2011) showed that although the extremes seen over a hemisphere, and particularly for the north Atlantic and north Pacific, in general show no significant signs of change in a warmer climate, there may be regional effects. Some regional studies also indicate that the intensity and duration of storms may increase, as indicated by e.g. Grabemann and Weisse (2008) for the North Sea, who found an increase in the 99th percentile wind speed over North Sea by 7 %. Debernard and Røed (2008) also found an increase in the 99th percentile wind speed from west of the British Isles and eastwards over the North Sea but only 2–4 %. They also indicated a reduction of the same parameter south of Iceland. Grabemann and Weisse (2008) and Debernard and Røed (2008) both studied changes in wave conditions and that will be referred to in more details in the next chapter. Projected increased storm activity over northern Europe is also reported by Ulbrich et al. (2008) and Donat et al. (2010).

Della-Marta and Pinto (2009) investigated the effects of climate change on the return period (RP) of the storm indicators minimum central pressure (CP) and

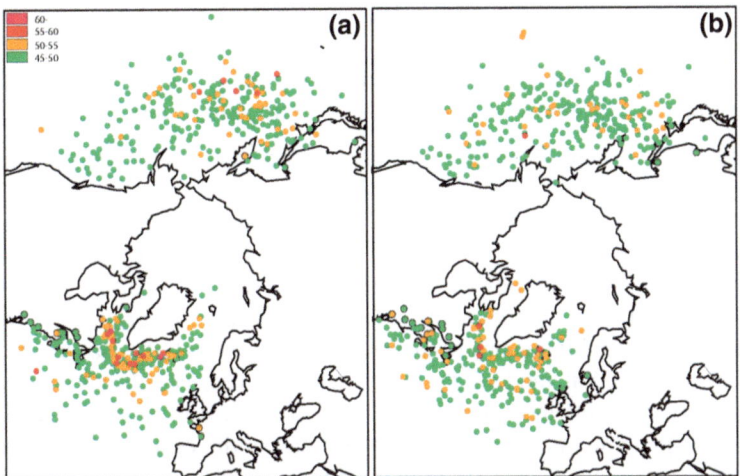

Fig. 2.5 Locations of storms on the northern hemisphere in winter (December, January and February) when they have their maximum wind speed at 925 hPa, for storms with maximum wind speed above 45 ms^{-1} for **a** the 20th century and **b** the 21st century. From Bengtsson et al. (2009)

maximum vorticity (VOR) in the North Atlantic. They used the ECHAM5 model with the emission scenarios A1B and A2 to compare the RPs of CP and VOR for the periods 1960–2000 and 2060–2100 using both a peak over threshold (POT) approach and the Generalized Pareto Distribution (GPD). Their results indicate that the RP of both the ocean basin minimum CP and maximum VOR remain unchanged through the 21st century. However, for maximum VOR the RP is shortened for the region British Isles—North Sea—western Europe already from 2040, indicating that the extremes expressed as e.g. the 50 year RP value will increase. They argue, though, that the estimated reduction in RPs for this region may be excessively large. The signal in changes of RP for CP is much weaker except for low intensity storms but this may be due to bias in the mean sea level pressure of the background field.

2.5.2 Tropical Cyclones

Concerning possible changes in tropical cyclone activity the Sixth WMO International Workshop on Tropical Cyclones (IWTC-VI) included the following points amongst its consensus statements in the proceedings (WMO 2007):

1. It is likely that some increase in tropical cyclone peak wind-speed and rainfall will occur if the climate continues to warm. Model studies and theory project a 3–5 % increase in wind-speed per degree Celsius increase of tropical sea surface temperatures.
2. Although recent climate model simulations project a decrease or no change in global tropical cyclone numbers in a warmer climate, there is low confidence in

this projection. In addition, it is unknown how tropical cyclone tracks or areas of impact will change in the future.
3. If the projected rise in sea level due to global warming occurs, then the vulnerability to tropical cyclone storm surge flooding would increase.

Again, these conclusions have been strengthened through the work of Knutson et al. (2010), who summarized progress since 2006 by stating that "future projections based on theory and high-resolution dynamical models consistently indicate that greenhouse warming will cause the globally averaged intensity of tropical cyclones to shift towards stronger storms, with intensity increases of 2–11 % by 2100. Existing modelling studies also consistently project decreases in the globally averaged frequency of tropical cyclones, by 6–34 %. Balanced against this, higher resolution modelling studies typically project substantial increases in the frequency of the most intense cyclones, and increases of the order of 20 % in the precipitation rate within 100 km of the storm centre. For all cyclone parameters, projected changes for individual basins show large variations between different modelling studies".

2.6 Expected Changes in Waves in the Twenty-first Century

The global climate models used in the AR4.1 did not calculate the wave conditions. Available studies use statistical relations between wave heights and sea level pressure (statistical downscaling) or the winds from the global models to run limited area wave models (dynamic downscaling) in order to predict the future wave climate. Again, one must be aware of several sources of uncertainty of wave climate projections, such as the emission scenario, the assumptions on which the global climate model is based, starting conditions for the global model, choice of model to generate the wave fields and choice of approach to extreme value analysis.

Most studies of the wave conditions in climate scenarios for the next decades are regional and focusing on the northern hemisphere, and in particular the North Atlantic. Two studies that took a global view are Wang and Swail (2006b) and Caires et al. (2006). Wang and Swail (2006b) showed the expected changes between the 1990s and the 2080s in the 20 year return values for significant wave in northern winter (January, February and March, JFM) and northern summer (July, August and September, JAS) as estimated for the A2 emission scenario. The values are averages over three models, one Canadian (CGCM2), one British (HadCM3) and one German (ECHAM4). Wang and Swail (2006b) indicated increase in the 20 year extreme values in the northern hemisphere and a general decrease in the southern hemisphere in January–March, except near the Antarctic in the Pacific Ocean. The increase over the 90 years reaches more than 0.5 m in the north Pacific and off the east coast of the United States. In the Norwegian Sea

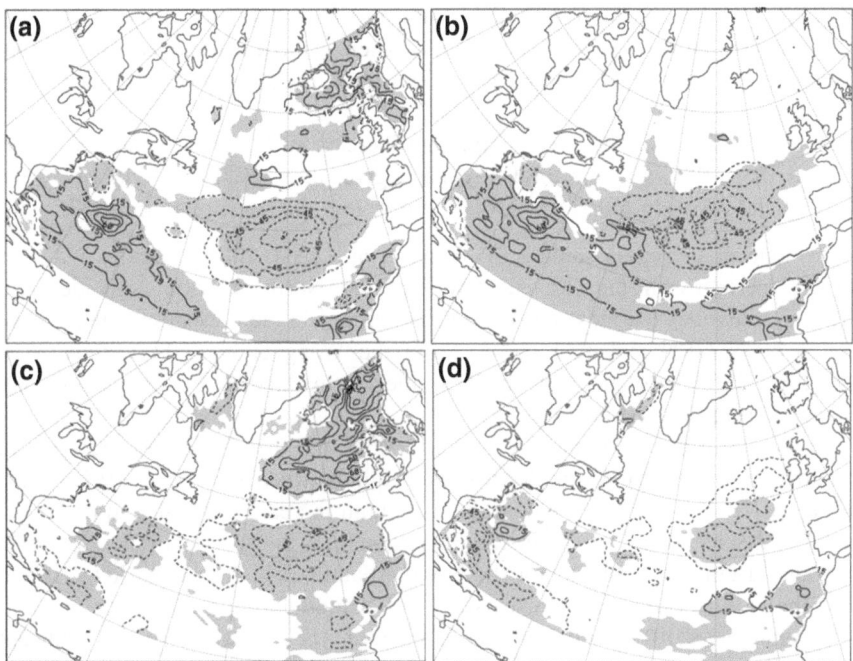

Fig. 2.6 Changes in the 20 year return values of SWH (H_{20yr}) from 1990 to 2080 (2080s minus 1990s), as projected in **a** winter with forcing scenario A2; **b** winter with forcing scenario B2; **c** fall with forcing scenario A2; **d** fall with forcing scenario B2. The contour interval is 15 cm. *Solid* and *dashed lines* indicate positive and negative contours, respectively. *Shading* indicates areas of significant quadratic trends in the location parameter of the SWH extremes. From Wang et al. (2004)

and off the northeast coast of Brazil the increase is 0.1–0.3 m, whereas the changes off West Africa are minor.

Wang et al. (2004) presented a more detailed analysis of future changes in wave conditions in the North Atlantic and Wang and Swail (2006a) extended this analysis to the include the North Pacific. In both cases the authors used only one climate model, the Canadian CGCM2, but now for three different emission scenarios: (1) the IPCC IS92a scenario; (2) the A2 scenario; and (3) the B2 scenario. For each of the scenarios both Wang and Swail (2006a) and Wang et al. (2004) presented fields of changes in seasonal mean, 90 % percentile and 20 year return period significant wave heights between 1990 and 2080 (2080s minus 1990s).

Wang et al. (2004) showed the effect of different emission scenarios on the 20 year return period significant wave height. In Fig. 2.6 (from Wang et al. 2004) we note the following:

- In the A2 winter (January–Febraury–March) scenario the largest increases are in a region eastwards of Florida, USA, and southeast of Iceland. The increase in both places is up to 0.6 m.

- In the B2 winter scenario the areas with maximum increase in the Norwegian Sea and the North Sea have disappeared, whereas the maximum eastwards from Florida has increased to around 1.0 m. In winter IS92a gives very similar results as B2.
- In fall the A2 scenario gives maximum increase in the Norwegian and North Seas, 0.6–1.0 m, whereas the B2 scenario gives only small differences between 1990 and 2080. In fall IS92a gives more or less the same picture as A2.

In none of the cases studied by Wang et al. (2004) do the regions with the largest increases in the 20 year SWH in the North Atlantic coincide with the region that has the most severe wave conditions in the present climate, i.e. the region 50°N–60°N and 10°E–30°E. From the Wang et al. (2004) study it is, therefore, difficult to judge if and how the North Atlantic extreme SWH will change during the 21st century.

In the North Pacific Wang and Swail (2006a) find that the largest increases in the 20-year return period significant wave height between 1990s and 2080s occurs in the eastern mid-latitudes, where the maximum increase is 1.0–1.3 m in the A2 and IS92a scenarios. This is outside the region with the most extremes waves in the present climate. Thus, as for the North Atlantic, it is difficult to judge if and how the North Pacific extreme sea state will change during the 21st century.

Mori et al. (2010) used the differences between the top ten values of significant wave height for two periods, January 1979–December 2003 and January 2075–December 2099 as a measure of climate change signal in the wave conditions. Using a 1.25° resolution global wave model driven by winds from the Japanese global circulation model MRI-JMA with the emission scenario A1B, they found significant increases in the average top ten SWH in the North Pacific and off the coast of Japan and only minor changes in the Indian, Antarctic and North Atlantic Oceans.

Regional studies strengthen the impression that the projected changes in extreme wave conditions are likely to be location dependent. Studies carried out for the North Atlantic and the North and Norwegian Sea, summarized below, confirm this.

Swail and Wang (2002) found that the 20 year return significant wave height in the North Sea would increase between 0.7 m and 1.15 m between the 1970s and the 2080s, depending on emission scenario.

Grabemann and Weisse (2008) found increases in the 99th percentile of the long-term significant wave height from present to the end of the 21st century to be 0.25–0.35 m as an average over four climate model/emission scenario combinations from present to the end of the 21st century. However, the range for the northern North Sea varied from −0.1 m to 0.6 m and the authors assign an uncertainty to the mean value of 0.6–0.7 m.

Debernard and Røed (2008) found that the annual 99-percentiles of significant wave height may increase 6–8 % along the North Sea east coast and in the Skagerrak, and 4 % or less in the North and Norwegian Seas and west of the British Isles by the end of the 21st century (2071–2100). The results indicate also more frequent strong wind events with higher extreme surge and wave events in the future. Another robust result is that the 99 % significant wave height southwest of Iceland will decrease by approximately 6 %. Despite the robustness of the mentioned results, the authors relate un-quantified uncertainty to these estimates due to imperfections of the analysis carried out.

2.7 Changes in Other Parameters that May Impact Ocean Transportation

2.7.1 Water Level

According to AR4.1 the global mean sea level rose by about 0.20 m between1870 and 1999. AR4.1 placed high confidence in the result. The rise was not constant, taking place at a rate of 1.8 ± 0.5 mm yr^{-1} over 1961–2003 and at a rate of 3.1 ± 0.7 mm yr^{-1} over 1993–2003. AR4.1 did not draw any conclusion of whether the faster rate of increase during the latter period reflected decadal variability or an increase in the longer term trend. However, it is not the mean sea level alone that may create the largest challenges for the coastal populations. Storm induced high water, the storm surges, are causing much damage to coastal infrastructure. Such surges may also change as a result of changing storm tracks, frequencies and/or intensities.

According to AR4.1 the rise in mean sea level and variations in regional climate are likely to have caused an increasing trend of extreme high water worldwide in the late 20th century and it was also found that human contribution to the trend in extreme high sea levels was more likely than not. Menendez and Woodworth (2010) used data from 258 tide gauges across the globe and found a trend in extreme sea levels globally. The trend was more pronounced after the 1970's, and it was consistent with trends in mean sea level. Menendez and Woodworth (2010) also found that subtraction from the extreme sea levels of the corresponding annual median sea level results in a reduction in the magnitude of trends at most stations, leading to the conclusion that much of the change in the extremes is due to change in the mean values. Studies at particular locations support this finding. Haigh et al. (2010) analysed sea level records from 18 tide gauges in the English Channel during the 20th century. They found that the extreme sea levels increased at all 18 sites but not at rates significantly different from the changes in the mean.

The AR4.1 projected sea level rise for 2090–2099 relative to 1980–1999 in the range 0.18–0.59 m, including the effects of ocean thermal expansion, glaciers and ice caps, across all scenarios. If allowance is made for a possible rapid dynamic response of the Greenland and West Antarctic ice sheets, an additional contribution to sea level rise was estimated to be 0.10–0.20 m by 2090–2099 using a simple linear relationship between sea level and projected temperature. Due to uncertainties caused by insufficient understanding of the dynamic response of ice sheets, AR4.1 also noted that a larger contribution could be possible. Indeed, later studies have indicated significantly higher sea level rise. Vermeer and Rahmstorf (2009) used a simple relationship between the global sea level variations on time scales of decades to centuries to global mean temperature and found that the relationship projects sea level rise between 0.75 and 1.90 m between 1990 and 2100, lowest for the B1 emission scenario and highest for the A1FI scenario.

Several studies have focused on regional changes in storm surges and extreme sea levels due to climate change. Debernard and Røed (2008) investigated storm surge in north European waters between the periods 1961–1990 and 2071–2100 using the same combinations of GCMs and emission scenarios as described above in Sect. 2.6. As for waves they found large differences between models but reported statistically significant changes of the 99th percentile. There was a decrease south of Iceland and an increase of 8–10 % along coast of the eastern coast of the North Sea and the northwest of the British Isles, mainly in the winter season. Haigh et al. (2011) reported that the exceedance frequency of extreme high sea levels along the south coast of UK would on average increase over the twenty-first century by a factor of 10, 100 and about 1800, respectively, for the low, medium and high emissions scenarios (these gave 12, 40 and 81 cm total ocean rise, respectively). Debernard and Røed (2008) pointed out that it is not only the wind speed and propagation direction that influence the storm surges but that also the propagation of the storm centre relative to the coast line will have impact on the response. This is supported by Sterl et al. (2009), who found no statistically significant increase in the 10000 years return period storm surge along the Dutch coast through the 21st century, despite projected changes in the wind speed. However, the future extreme winds were not surge-generating northerlies but rather non-surge generating southerlies. Sterl et al. (2009) used only one model and one emission scenario, and a larger ensemble of models and emission scenarios is needed to confirm the results.

Debernard and Røed (2008) also mentioned that storm surges usually propagate as trapped planetary waves and that a single event may affect a large area. The local impact, though, will depend on the topography and coastline geometry.

Thus, presently and based on the above referenced publications, we conclude that it is likely that climate change have resulted in sea level rise and that the rise may continue in a future with increased emissions of greenhouse gases. However, it is not possible to quantify increase of extreme sea levels and storm surges in general at present; rather, changes will by dependent on a location.

2.7.2 Sea Ice

Sea ice in the Arctic has shown dramatic changes over the last 30 years. The extent of summer ice (September) declined by 8.9 % per decade between 1979 and 2009 and the winter ice (March) by 2.5 % per decade. September of 2007 and 2011 had both record low ice extent with more ice in the years 2008–2010 (at the time of writing 2012 have less ice than 2007 at the end of August). However, it is likely that the total Arctic sea ice volume has declined continuously, as the amount of multi-year ice has decreased and therefore the mean ice thickness.

It is expected that increase of the average Earth surface temperature will be twice as high in the Arctic compared to other parts of the Earth. Consequently, the sea ice cover is expected to be reduced significantly. Stroeve et al. (2007) showed how a range of climate models project future September ice extent in the Arctic Ocean, along with observed ice extent up until 2006. They show that the ice cover is retreating faster than the models predict. After 2006 the retreat has been even faster. The models indicate that the ice cover in summer may practically disappear around 2050 (e.g. Wang and Overland 2009); before 2020 has even been suggested by some investigations (e.g. Maslowski 2008). An ice free Arctic winter is not predicted by any model, but the winter ice may be limited to first year ice and, therefore a likely maximum thickness of 2.0–2.5 m.

For updated information on the sea ice conditions, see e.g. the web site of the US National Snow & Ice Data Center (http://nsidc.org/arcticseaicenews/) and the Arctic Regional Ocean Observing System (Arctic ROOS) (http://arctic-roos.org/).

It is outside the scope for this monograph to look at the implications a diminishing ice cover may have on marine transport. Several publications have discussed possible implications, the most extensive one being the Arctic Marine Shipping Assessment Report (AMSA 2009). Peters et al. (2011) looked at possible increases in container traffic across the Arctic Ocean and give further references on the subject.

References

Arctic Council (2009) Arctic marine shipping assessment (AMSA) 2009 Report, April 2009

Alexandersson H, Tuomenvirta H, Schmith T, Iden K (2000) Trends of Storms in NW Europe Derived from an Updated Pressure Data Set. Clim Res 14:71–73

Bengtsson L, Hodges K.I, Keenlyside N (2009) Will extra-tropical storms intensify in a warmer climate? J Clim 2276–2301

Bitner-Gregersen E.M, Hagen, Ø (1990) Uncertainties of data for the offshore environment. J Struct Saf 7

Bärring L, Fortuniak K (2009) Multi-indices analysis of southern scandinavian storminess 1780–2005 and links to interdecadal variations in the NW Europe–North sea region. Int J Climatol 29:373–384

Bärring L, Storch HV (2004) Scandinavian storminess since about 1800. Geophys Res Lett Vol. 31 L20202, doi:10.1029/2004GL020441

Caires S, Swail VR (2004) Global wave climate trend and variability analysis. In: 8th International workshop on wave hindcasting and forecasting, November 14–19 North Shore, Oahu

Caires S, Swail VR, Wang XL (2006) Projection and analysis of extreme wave climate. J Clim 19:5581–5605

Cartwright DE, Longuet-Higgins MS (1956) Statistical Distribution of the maxima of a random function. Proc Roy Soc A 237:212–232

Champion AJ, Hodges KI, Bengtsson L, Keenlyside NS, Esch M (2011) Impact of Increasing Resolution and a Warmer Climate on Extreme Weather from Northern Hemisphere Extratropical Cyclones. Tellus 63A:893–906

Debernard JB, Røed LP (2008). Future wind, wave and storm surge climate in the Northern Seas: a Revisit. Tellus 60A:427–438

Della-Marta PM, Pinto JG (2009) Statistical uncertainty of changes in winter storms over the North Atlantic and Europe in an ensemble of transient climate simulations. Geophys Res Lett 36:L14703

Donat MG, Leckebusch GC, Pinto JG, Ulbrich U (2010) European storminess and associated circulation weather types: future changes deduced from a multi-model ensemble of gcm simulations. Geophys Res Lett 28:195–198

Grabemann I, Weisse R (2008) Climate change impact on extreme wave conditions in the north sea: an ensemble study. Ocean Dyn 58:199–212

Gulev SK, Grigorieva V (2004) Last century changes in ocean wind wave height from global visual wave data. Geophys Res Lett 31:L24302. doi:10.1029/2004GL021040

Gulev SK, Gregovieva V (2006) Variability of the winter wind waves and swell in the North Atlantic observing ship data. J Clim 19:5667–5685

Haigh I, Nicholls R, Wells N (2010) Assessing changes in extreme sea levels: Application to the English Channel, 1900–2006. Cont Shelf Res 30(9):1042–1055

Haigh I, Nicholls R, Wells N (2011) Rising sea levels in the english channel 1900 to 2100. Proceedings of the ICE - Maritime Engineering 164(2):81–92. doi:10.1680/maen.2011.164.2.81

IPCC (Intergovernmental Panel on Climate Change) (2000) Special report on emission scenarios. (Nakicenovic N, Nebojsa Nakicenovic, Alcamo J, Davis G, de Vries B, Fenhann J, Gaffin S, Gregory K, Griibler A, Jung TY, Kram T, La Rovere EL, Michaelis L, Mori S, Morita T, Pepper W, Pitcher H, Price L, Riahi K, Roehrl A, Rogner HH, Sankovski A, Schlesinger M, Shukla P, Smith S, Swart R, van Rooijen S, Victor N, DadiZ). Cambridge University Press, Cambridge, United Kingdom and New York

IPCC (2001) Climate change (2001) The scientific basis. Contribution of Working Group I to the third assessment report of the intergovernmental panel on climate change. Houghton, JT,Y Ding, DJ Griggs, M Noguer, PJ van der Linden, X Dai, K Maskell, and CA Johnson (eds.). Cambridge University Press, Cambridge, United Kingdom and New York, NY, USA, p 881

IPCC (2007a) Climate change 2007. Synthesis report. Contribution of working grpous I, II and III to the fourth assessement report of the intergovernmental panel on climate change. Core Writing team Pachauri RK, Reisinger A (eds). IPCC Geneva, Switzerland p 104

IPCC (2007b) Climate Change (2007). The physical science basis. Contribution of working group i to the fourth assessment report of the intergovernmental panel on climate change Solomon, S., D. Qin, M. Manning, Z. Chen, M. Marquis, K.B. Averyt, M. Tignor and H.L. Miller (eds.). Cambridge University Press, Cambridge, United Kingdom and New York, NY, USA, 996

IPCC (2011). Summary for policymakers. In: Intergovernmental panel on climate change special report on managing the risks of extreme events and disasters to advance climate change adaptation. Field CB, Barros V, Stocker TF, Qin D, Dokken D, Ebi KL, Mastrandrea MD, Mach KJ, Plattner GK, Allen S, Tignor M, Midgley PM (eds) Cambridge University Press, Cambridge, United Kingdom and New York, NY, USA

IPCC (2012) Managing the risks of extreme events and disasters to advance climate change adaptation. A special report of working groups I and II of the intergovernmental panel on climate change Field CB, V Barros, Stocker TF, Qin D, Dokken DJ, Ebi KL, Mastrandrea MD, Mach KJ, Plattner G-K, Allen SK, Tignor M, Midgley PM (eds). Cambridge University Press, Cambridge, UK, and New York, NY, USA, p 582

Knutson TR, McBrite JL, Chan J, Emanuel K, Holland G, Landsen C, Held I, Kissing JP, Srivastava AK, Seegi M (2010) Tropical Cyclones and climate change. Nat Geosci 3:157–163

Longuet-Higgins MS (1952) On statistical distribution of the heights of sea waves. J Marine Res Vol. XI, No. 3

Maslowski W (2008) When will summer Arctic sea ice disappear. Symposium on drastic change in the earth system during global warming. Sapporo, Japan

Menendez M, Woodworth PL (2010) Changes in extreme high water levels based on a quasi-global tide-gauge dataset. J Geophys Res 115(C10011)

Mori N, Yasuda T, Mase H, Tom T, Oku Y (2010) Projection of extreme wave climate change under global warming. Hydrological Res Lett 4:15–19

Peters GP, Nilssen TB, Lindholt L, Eide MS, Glomsrød S, Eide LI, Fuglestvedt JS (2011) Future emissions from oil, gas, and shipping activities in the Arctic. Atmos Chem Phys Discuss 11:4913–4951 www.atmos-chem-phys-discuss.net/11/4913/2011/ doi:10.5194/acpd-11-4913-2011

Pfizenmayer, A. and H.v. Storch (2001). Anthropogenic climate change shown by local wave conditions in the North Sea. Clim Res 19:15–23

Sterl A, van den Brink H, de Vries H, Haarsma R, van Meijgaard E (2009) An Ensemble Study of Extreme Storm Surge Related Water Levels in the North Sea in a Changing Climate. Ocean Sci 5(3):369–378

Stroeve J, Holland MM, Meier W, Scambos T, Serreze M (2007) Arctic sea ice decline: faster than forecast. Geophysical Res Lett 34:L09501. doi:10.1029/2007GL029703,2007

Swail VR, Wang XL (2002). The wave climate of the North Atlantic – Past, present and future. 7th international workshop on wave hindcasting and forecasting, October 21–25, Banff, Alberta, Canada

Ulbrich U, Pinto JG, Kupfer H, Leckebusch GC, Spangehl T, Reyers M (2008) Changing northern hemisphere 6 storm tracks in an ensemble of ipcc climate change simulations. J Clim 21(8):1669–1679

Vermeer M, Rahmstorf S (2009) Global Sea Level 1 Linked to Global Temperature. Proc Natl Acad Sci USA 106(51):21527–21532

Wang M, Overland JE (2009) A sea ice free summer Arctic within 30 years? Geophys Res Lett 36:L07502. doi:10.1029/2009GL037820

Wang XL, Swail VR (2006a). Historical and possible future changes of wave heights in northern hemisphere ocean. Atmosphere-Ocean Interactions Perrie W (ed.) Vol 2. Wessex Institute of Technology Press, Southampton 240

Wang XL, Swail VR (2006b) Climate change signal and uncertainty in projections of ocean wave heights. Clim Dyn 26:109–126. doi:10.1007/s00382-005-0080-x

Wang XL, Zwiers FW, Swail VR (2004) North Atlantic ocean wave climate change scenarios for the twenty-first century. J Clim 17:2368–2383

Wang XL, Zwiers FW, Swail VR, Feng Y (2009a) Trends and variability of storminess in the Northeast Atlantic region, 1874–2007. Clim Dyn 33(7–8):1179–1195

Wang XL, Swail VR, Zwiers FW, Zhang X, Feng Y (2009b) Detection of external influence on trends of atmospheric storminess and northern oceans wave heights. Clim Dyn 32(2–3):189–203

Wasa Group (1998) Changing waves and storms in the northeast atlantic? Bull Am Meterological Soc 79:5

WMO (World meteorological Organization) (2007). World Weather Research Programme WWRP 2007-1 Sixth WMO International Workshop on Tropical Cyclones (IWTC-VI) (San Jose, Costa Rica, 21–30 November 2006). WMO TD No. 1383

Worley SJ et al (2005) ICOADS release 2.1 data and products. Int J Climatology 25:823–842

Young RI, Zieger S, Babanin AV (2011) Global trends in wind speed and wave height. Science 332:451–455

Chapter 3
Uncertainties

3.1 General

The oceanographic community has always been concerned with providing environmental models and data which approximate the physics of the ocean in the most accurate way. Industry, on the other hand, needs accurate data and models for design purposes. Although uncertainties of data and models were discussed before the 1980s, they were not systematically quantified. Further development of the reliability methods (Madsen et al. 1986) and their implementation by some parts of the industry in the 1980s has brought much focus onto the uncertainties associated with environmental description. The PROBabilistic Analysis program PROBAN® developed by Det Norske Veritas at the end of the 1980s, and continuously improved since then (DNV 2002), is still one of the leading software packages for reliability calculations and is used by academia as well as industry. Reliability methods allow quantification, in a probabilistic way, of the uncertainties in the different parameters that govern structural integrity.

3.2 Definition of Uncertainties

Bitner-Gregersen and Hagen (1990) suggested classification of uncertainties for environmental description. The proposed definitions were later generalised and in 1992 included in DNV Rules (DNV 1992).

Generally, uncertainty related to an environmental description may be divided into two groups: aleatory (natural variability) uncertainty and epistemic (knowledge) uncertainty. Aleatory uncertainty represents a natural randomness of a quantity, also known as intrinsic or inherent uncertainty, e.g. the variability in wave height over time. Aleatory uncertainty cannot be reduced or eliminated (see Skjong et al. 1995).

Epistemic (knowledge) uncertainty represents errors which can be reduced by collecting more information about a considered quantity and improving the

E. M. Bitner-Gregersen et al., *Ship and Offshore Structure Design in Climate Change Perspective*, SpringerBriefs in Climate Studies,
DOI: 10.1007/978-3-642-34138-0_3, The Author(s) 2013

methods of measuring it. In accordance with Bitner-Gregersen and Hagen (1990), this uncertainty may be classified into: data uncertainty, statistical uncertainty, model uncertainty and climatic uncertainty.

1. Data uncertainty is due to imperfection of an instrument used to measure a quantity, and/or a model used for generating data. If a quantity considered is not obtained directly from the measurements but via some estimation process, e.g. significant wave height, then the measurement uncertainty must be combined with the estimation or model uncertainty by appropriate means.
2. Statistical uncertainty, often referred to as estimation uncertainty is due to limited information such as a limited number of observations of a quantity (sampling variability) and is also due to the estimation technique applied for evaluation of the distribution parameters. The latter can be regarded as the model uncertainty.
3. Model uncertainty is due to imperfections and idealisations made in physical process formulations as well as in choices of probability distribution types for representation of uncertainties.
4. Climatic uncertainty addresses the representativeness of measured or simulated wave history for the (future) time period and area for which design conditions need to be provided. A data set has to be sufficiently long to eliminate climatic uncertainty, e.g. to avoid biasing towards years characterized by severe winds or by calm weather only.

To characterise the accuracy of a quantity, e.g. significant wave height SWH, it is necessary to distinguish systematic error (bias) and precision (random error) with reference to the true value τ, which usually is unknown.

Generally, environmental description will be affected by all types of epistemic uncertainties to varying degrees. Identification of uncertainties and their quantification represents important information for risk assessment in design and operation of marine structures. High uncertainty of environmental description may lead to over-design or under-design of marine structures, with significant economic/risk impact. Several authors have demonstrated in the past the importance of uncertainties for calculations of load and responses. Offshore industry had a leading role here. The shipping industry has tended to lag behind the offshore industry in these investigations. In the last decade also the shipping industry, as well as academia, has focused increasingly on studying sensitivity of ship load and responses to adopted uncertainties, see Bitner-Gregersen et al. (2002), Hørte et al. (2007), Bitner-Gregersen and Skjong (2008), Nielsen et al. (2009).

Enhancing safety at sea through specification of uncertainties related to environmental description is today one of the main concerns of the shipping industry in general and the Classification Societies in particular. The offshore industry is also much concerned with it.

Specification of uncertainties for environmental description is not an easy task because the true value τ is usually unknown and needs to be estimated based on available information. For the integrated wave parameters, for example, the values provided by wave rider buoys are commonly adopted as the true values.

The situation is even more difficult for environmental models where experimental tests or the average values of recognized models are used as the reference values today. Further discussion on how to estimate the true value τ is still called for.

3.3 Uncertainties Related to Wave Climate Projections

Identification of uncertainties specified in Sect. 3.2 is of importance for design, particularly as no field observations will be available for validation of the projected future climate.

Specification of uncertainties is also of much concern to climate change researchers because they influence the climate model's simulation of past, contemporary and future climate. Therefore much effort goes into reducing the uncertainties. The topic was discussed by the Workshop on Climate Change organized by the WMO and the OGP in Geneva in 27–29 May 2008.

Projections of wind and wave conditions for the twenty-first century will be subject to all four types of uncertainties described in the preceding chapter, particularly:

- Model uncertainty. There are at least a score of Global Circulation Models (GCM) in use around the world. They may give substantially different results, as described by Covey et al. (2003), who also state that the differences are less than what would have been anticipated from earlier studies. An updated intercomparison is due in 2013. The GCM models provide input to Regional Circulation Model (RCM) models. The degree to which GCM and RCM models have sufficient resolution and/or internal physics to realistically capture the meteorological forcing responsible for changes in met-ocean conditions is regionally dependent. For example current GCMs are unable to realistically represent tropical cyclones.
- Data uncertainty relates both to data used to describe present day wind and to wave data generated by wave models. The hindcasts include assimilated satellite data calibrated towards wave buoy data, and are often called "corrected hindcasts". They will be affected by uncertainty of data used for calibration of satellite data as well as the assumptions on which the wave model is based on.
- Climate uncertainty, i.e. the forcing data used to drive the GCM. They vary with respect to assumed economic and societal global development and, therefore, future emission levels. They will also be affected by the projected time period.

The papers and studies referenced in Sect. 2.6 all indicate that a changing climate will have impact on extreme wave conditions in the world oceans but there are differences in the projections and even in-consistencies between the studies. This is not surprising given that the results were obtained using both different climate models and different climate forcing (emission scenarios).

As mentioned above in addition to the uncertainties introduced by choice of GCM climate model and climate forcing there are other factors related to regional modelling that will also add to the uncertainties when considering projections of future extreme wave conditions. Ocean wave heights are not directly available from global climate models and projections have to be made through some form of "downscaling" approaches of either statistical or dynamical character. The statistical approaches imply establishing statistical relations between output from the global climate models, e.g. gradients of sea level pressure (SLP) and observed or hindcast wave height parameters. Dynamical downscaling involves use of regional or local ocean wave models driven by the output from the global or regional climate models, e.g. surface wind speed and direction. Thus the choice of statistical vs. dynamical downscaling as well as the choice of statistical and dynamical wave models will introduce uncertainties in the projected future wave conditions.

Another factor that adds uncertainty is the way extremes are defined and estimated. With extremes we here understand the K-year return period value. Thus the choice of extreme value estimation may influence the results.

None of the reviewed papers have a systematic evaluation of the contribution of all the factors that influence the total uncertainty in projections of future wave conditions. Therefore it is difficult to quantitatively assess the relative contribution between more than two factors at a time.

3.3.1 Effect of Climate Model and Climate Forcing

Wang et al. (2004), Wang and Swail (2006a) as well as Caires et al. (2006) used one climate model (the Canadian CGCM2) for different emission scenarios. Wang and Swail (2006b), however, represented the average of three models for the emission scenarios A2 and B2. Their results confirm the qualitative picture and patterns of the other studies using only the CGCM2, but the magnitude of changes were generally smaller, indicating an influence of the choice of climate model on the end results. Indeed, Wang and Swail (2006b) found that the choice of climate model contributes most to wave projection uncertainty. This conclusion seems to be supported by Debernard and Røed (2008). Grabemann and Weisse (2008) also stated that there is a large uncertainty related to differences among climate models and found these to be larger than those related to differences among the climate forcing, or emission scenarios.

It is, however, important to note that the above studies mainly used the emission scenarios A2 and B2 (IS92a that was used in some cases is very similar to A2). These two scenarios do not represent the full variability of the IPCC scenarios and the results would probably have looked different if also the more extreme scenarios A1 and B1 were included (IPCC 2000). Both Debernard and Røed (2008) and Mori et al. (2010) used the A1B but combined with only one model, the BCCR model of the Bjerknes Center for Climate Research and the Japanese MRI-JMA, respectively. However A1B is an intermediate between A2 and B2 and closest to B2 and it is doubtful if this would shed more light on the contribution of different scenarios to the uncertainty.

It is difficult to quantify the uncertainties caused by different combinations of climate model and forcing scenario. Debernard and Røed (2008) combined the A2 and B2 scenarios with the Hadley Centre model HADAM3H, denoted as HADA2 and HADB2, and the Max Planck Institute model ECHAM4 with B2, denoted MPIB2, in addition to the combination of the BCCR model with the A1B scenario. Although they find some robust results, like the decrease in wave conditions southwest of Iceland and increase in the North and Norwegian Seas there are also indications that the uncertainties caused by different combinations may be of the same order as the projected change itself. Debernard and Røed (2008) showed quantile (qq) plots of scenario vs. control run and the relative increases of the ten largest sea states for four locations. E.g. for the K4 buoy location near Rockall west of Ireland, see map in Fig. 3.1, both the MPIB2 and HADB2 scenarios show changes in the three largest sea states of 7–8 % but with opposite signs (MPIB2 shows a decrease and HADB2 an increase). Different signs of the changes are also found at location M (Mike Station) in the Norwegian Sea and Tromsøflaket in the Barents Sea, although not as pronounced and consistent as for the K4 location. For Ekofisk in the North Sea the two scenarios agreed better.

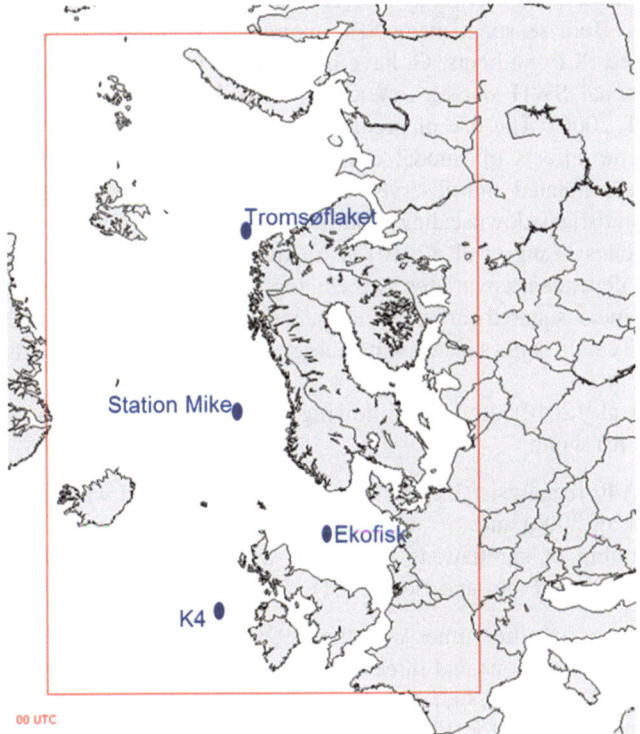

Fig. 3.1 The Norwegian and North Sea with some of the locations referred to by Debernard and RØed (2008). Red rectangle is the domain of the hindcast model NORA10 by the Norwegian Meteorological Institute (met.no)

The results of Debernard and Røed (2008) are very similar to those of Grabemann and Weisse (2008), who considered the combinations of the Hadley Centre and Max Planck Institute climate models with the A2 and B2 forcing scenarios in the North Sea south of 57° 45'. Grabemann and Weisse (2008) found uncertainties in the 99th percentile significant wave height related to model choice to be up 0.7 m in the north, close to the coast of Norway. For differences between forcing scenarios the difference was up to 0.1 m northwest of Denmark. Both differences are of the same order as the projected changes.

3.3.2 Effect of Downscaling

Wang et al. (2010) compared dynamical and statistical downscaling methods for estimating seasonal statistics of SWH. The different downscaling approaches were valuated against the ERA40 wave data in terms of climatological characteristics and inter-annual variability. They considered only the North Atlantic.

Statistical downscaling methods for estimating SWH will generally establish statistical relationship between observed atmospheric predictors and observed SWH fields. Both seasonal mean SLP anomalies, P, and anomalies of seasonal mean squared SLP gradients, G, have been shown to be good predictors for predicting seasonal SWH statistics (Wang and Swail 2006a, b; Wang et al. 2004; Caires et al. 2006). The use of predictor anomalies, instead of predictor values, diminishes the effects of 'model climate biases', i.e. the systematic difference between the simulated and observed climate (long-term mean) fields. In order to make the statistical downscaling results more comparable with the wind-based dynamical ones Wang et al. (2010) also explored the potential of using surface wind speed dependent covariates as predictors for SWH by considering anomalies of seasonal mean squared surface wind speeds, W, as a predictor. Similar to G, this wind quantity represents surface wind energy. Use of W as predictors had not been explored before.

Wang et al. (2010) used the following global datasets of mean SLP, and of surface (10 m) wind:

1. The ERA40 reanalysis (Uppala et al. 2005), with global wave reanalysis (e.g. Caires et al. 2006) and
2. An ensemble of simulations conducted with the Canadian coupled climate model CGCM2 (Flato and Boer 2001).

CGCM2 was run three times with the IS92a forcing scenario, each run having different initial conditions and three datasets were stored: 1975–1994, 2040–2059 and 2080–2099. Only the first and third were used by Wang et al. (2010).

The ERA40 dataset for 1975–1994 was used as "observations" from which Wang et al. (2010) could estimate the bias of CGCM2 as well as obtaining standardized values of the predictors P (seasonal mean sea level pressure anomalies), G (anomalies of seasonal mean squared SLP gradients) and W (seasonal mean squared surface wind speed). The predictors were standardized using observed mean and standard deviations for the period 1975–1994.

Using 12-hourly values of the predictors and the predicted SWH, they found that it is sufficient to use W as predictor. In addition to the 12-hourly values for predictors they also used seasonal data. Thus there were four combinations—12-hourly and seasonal data each with non-standardized and standardized predictors. The standardized estimates were judged to be best and thus preferred.

For dynamical downscaling Wang et al. (2010) obtained global 6-hourly instantaneous wave heights by running the ODGP-2G wave model (Cox and Swail 2001) forced by the global 6 - or 12-hourly surface wind fields from the afore-mentioned CGCM2 ensemble of three runs. In addition the model was also run using winds that were adjusted for model wind climate biases. Inter-comparison of runs with and without adjusted wind climate estimates suggests that the former estimates are better than the latter because the latter suffers from the effects of model wind climate biases.

For the statistical downscaling Wang et al. (2010) concentrated on projections of changes in mean and 20-year return period values between 1975–1994 and 2080–2099 using both standardized 12-hourly and standardized seasonal data with seasonal mean squared surface wind speed, W, as predictor. For the dynamical downscaling they consider the same periods with adjusted wind climate.

The broad picture is that the dynamical and both statistical projections show similar patterns of projected changes in both seasonal means and extremes, in winter (January, February and March) as well as in fall (October, November and December). In winter, this pattern is characterized by increases in the eastern and western subtropical North Atlantic and decreases almost everywhere else. In the fall, it is characterized by increases west of British Isles and eastern subtropical North Atlantic, with decreases in the other areas.

On a smaller scale, there are some differences among the three sets of projections. The dynamical downscaled projections with adjusted wind climate show generally more extensive areas of large decreases than either of the statistical projections, especially for seasonal mean SWH. For a small region off Labrador in winter, the same dynamically downscaled projections show large increases, but the standardized statistically downscaled projections based on seasonal data sets show small decreases, in both seasonal means and extremes of SWH. Projections based on 12-hourly data sets show increases in seasonal extremes with small decreases in seasonal means.

In autumn, the mid latitude area of increase is located more eastward in the adjusted dynamical projections than in either of the statistical projections.

In some areas the difference between the different downscaling approaches for projected changes in the 20-year return period SWH from 1975–1994 to 2080–2099 are of the same order as the projected change itself. The projected changes found by Wang et al. (2010) appear to be 0.5–1.0 m of either sign.

To further illustrate the effect of dynamical downscaling we have used wave data from the ERA-Interim reanalysis (Dee et al. 2011) and the NORA10 hindcast data of the Norwegian Meteorological Institute (Reistad et al. 2011).

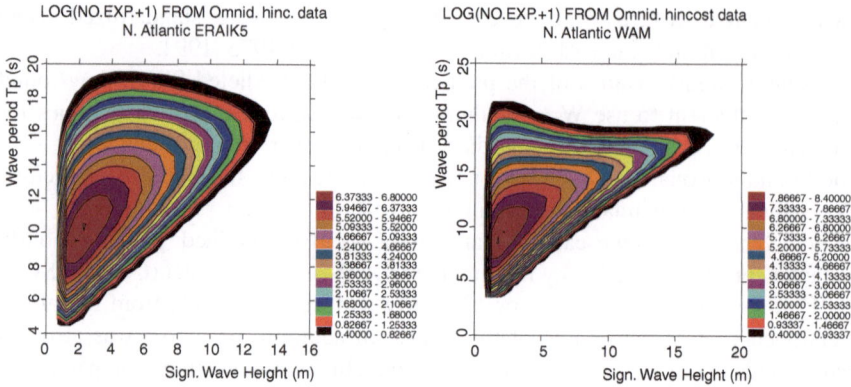

Fig. 3.2 Fitted joint probability densities of SWH and spectral wave period to ERA-Interim **a** and NORA10 data **b**

The horizontal resolution of the wave data of ERA-Interim is 110 km, whereas it is 10 km in NORA10. Scatter diagrams of significant wave height and spectral wave period have been established and fitted by a joint model, with a 3-parameter Weibull distribution for the SWH and a conditional lognormal distribution for the spectral wave period, T_p, see Bitner-Gregersen (1988), Sect. 6.4. and Eqs. (9–12). The fitted distributions are shown in Fig. 3.2.

The differences between two data sets are obvious and significant. The fitted probability densities differ not only in shape but also in extremes. The higher resolution NORA10 reveal more details and also higher waves. In fact, the difference in the 100-year significant wave height from the two models is 3–4 m.

3.3.3 Effect of Extreme Value Model

For design of ships, and marine structures in general, the extreme values of wind and wave parameters need to be estimated. Today, time-independent statistics are used in design. For climate change projections the non-stationary character of the climate, in terms of both climate change trends and natural variability cycles, needs to be taken into account.

Some attempt has been made to model non-stationary character of wind and wave climate. Caires et al. (2006) used the non-homogeneous Poisson process (NPP) to model extreme values of the 40-year long ECMWF Re-Analysis (ERA-40) data set for significant wave height. The model parameters were expressed as functions of the seasonal mean sea pressure anomaly and seasonal squared sea pressure gradients index. Using three scenarios, projections of the parameters of the non-homogeneous Poisson process were made; trends to these projections were determined and return-value estimates of the significant wave height up to the end of the twenty-first century were calculated. Estimates of the uncertainty resulting from the choice of non-stationary model for extremes were obtained by comparing estimates associated with the NPP estimates and the homologous estimates using a non-stationary generalized extreme value (NS-GEV) model. Caires et al. (2006) found that the results from the two statistical models are compatible by predicting the same type of trends and that estimates at a fixed time point have intersecting 95 % confidence intervals. They also found that the width of the confidence intervals can be quite large and larger than the projected changes, especially in the case of NS-GEV. The effects of choice of non-stationary model for extremes is largest where the confidence intervals are large.

Alternatively to climate models the Bayesian hierarchical models in space and time provide very valuable contribution to predictions of future met-ocean climate. As demonstrated for SWH by Vanem et al. (2011) the Bayesian models allow predicting climate changes in met-ocean conditions based on historical data and forcing scenarios, thus avoiding running climate models. Vanem and Bitner-Gregersen (2012) have applied successfully the Bayesian hierarchical model of Vanem et al. (2011) to monthly maxima at one location in the North Atlantic and predicted increase of extreme SWH. The model has been recently extended to account for different forcing scenarios, Vanem (2012). The obtained increase in the SWH monthly maxima is consistent with the results obtained by use of GCM and downscaling techniques.

References

Bitner-Gregersen EM (1988) Joint environmental model. Annex A, Det norske veritas report no. 87–31, Høvik: Madsen HO, Rooney P, Bitner-Gregersen E Probabilistic calculation of design criteria for ultimate tether capacity of snorre TLP

Bitner-Gregersen EM, Hagen O (1990) Uncertainties of data for the offshore environment. J Struct Saf 7:11–34

Bitner-Gregersen EM, Hovem L, Skjong R (2002) Implicit reliability of ship structures. Proceedings OMAE2002, Oslo, 23–28 June 2002

Bitner-Gregersen EM, Skjong R (2008) Concept for a risk based navigation decision assistant. Marine Struct 22(2009):275–286

Caires S, Swail VR, Wang XL (2006) Projection and analysis of extreme wave climate. J Climate 19:5581–5605

Covey C, AchutaRao KM, Cubasch U, Jones P, Lambert SJ, Mann ME, Phillips TJ, Taylor KE (2003) An Overview of Results from the Coupled Model Intercomparison Project (CMIP). Global Planet Change 37:103–133

Cox AT, Swail VR (2001) A global wave hindcast over the period 1958–1997: validation and climate assessment. J Geophys Res 106(C2):2313–2329

Debernard JB, Røed LP (2008) Future wind, wave and storm surge climate in the northern seas: a Revisit. Tellus 60A:427–438

Dee DP, Uppala SM, Simmons AJ, Berrisford P, Poli P, Kobayashi S, Andrae U, Balmaseda MA, Balsamo G, Bauer P, Bechtold P, Beljaars AC M, van de Berg L, Bidlot J, Bormann N, Delsol

C, Dragani R, Fuentes M, Geer AJ, Haimberger L, Healy SB, Hersbach H, H'olm EV, Isaksen L, Kållberg P, Köhler M, Matricardi M, McNally AP, Monge-Sanz BM, Morcrette J-J, Park B-K, Peubey C, de Rosnay P, Tavolato C, Th'epaut J-N, Vitart F (2011) The ERA-interim reanalysis: configuration and performance of the data assimilation system. Q J R Meteorol Soc 137:553–597

DNV (1992). Classification note 30.6: Structural reliability analysis of marine structures. July 1992

DNV (2002). PROBAN theory, general purpose probabilistic analysis program, the author L. Tvedt, Version 4.4, Høvik, Norway

Flato GM, Boer GJ (2001) Warming asymmetry in climate change simulations. Geophys Res Lett 28(1):195. doi:10.1029/2000GL012121

Grabemann I, Weisse R (2008) Climate change impact on extreme wave conditions in the North Sea: an ensemble study. Ocean Dyn 58:199–212

Hørte T, Skjong R, Friis-Hansen P, Teixeira AP, Viejo de Francisco F (2007) Probabilistic methods applied to structural design and rule development. Proceedings of the RINA conference development of classification & international regulations, Jan 2007, London, pp 24–25

Madsen HO, Krenk S, Lind NC (1986) Methods of structural safety. Prentice-Hall, Enlewood Cliffs 07632

Mori N, Yasuda T, Mase H, Tom T, Oku Y (2010) Projection of extreme wave climate change under global warming. Hydrol Res Lett 4:15–19

Nielsen UD, Friis-Hansen P, Jensen JJ (2009) A step towards risk-based decision support for ships-evaluation of limit states using parallel system analysis. Marine struct 22:660–669

Reistad M, Breivik O, Haakenstad H, Aarnes OJ, Furevik BR (2011) A high-resolution hindcast of wind and waves for the North Sea, the Norwegian Sea and the Barents Sea. J Geophys Res Oceans 116. doi:10.1029/2010JC006402. C05019

Skjong R, Bitner-Gregersen EM, Cramer E, Croker A, Hagen Ø, Korneliussen G, Lacasse S, Lotsberg I, Nadim F, Ronold KO (1995). Guidelines for offshore structural reliability analysis—General. DNV report no 95–2018

Vanem E, Huseby AB, Natvig B (2011) A Bayesian hierarchical spatio-temporal model for significant wave height in the North Atlantic. Stochastic environmental research and risk assessment, 29 Sept 2011, pp 1–24. doi:10.1007/s00477-011-0522-4

Vanem E, Bitner-Gregersen EM (2012) Stochastic modelling of long-term trends in the wave climate and its potential impact on ship structural loads. Appl Ocean Res 37(2012):235–248

Vanem E (2012). A stochastic model for long-term trends in significant wave height with a CO2 regression component. Proceedings of OMAE 2012 confernece, 1–6 July 2012, Rio de Janeiro, Brazil

Wang XL, Swail VR (2006a) Historical and possible future changes of wave heights in Northern Hemisphere Ocean. In: Perrie W (ed.) Atmosphere-Ocean interactions, vol 2. Wessex Institute of Technology Press, Southampton, p 240

Wang XL, Swail VR (2006b) Climate change signal and uncertainty in projections of Ocean wave heights. Clim Dyn 26:109–126. doi:10.1007/s00382-005-0080-x

Wang XL, Zwiers FW, Swail VR (2004) North Atlantic Ocean wave climate change scenarios for the twenty-first century. J Clim 17:2368–2383

Wang XL, Swail VR, Cox A (2010) Dynamical versus statistical downscaling methods for Ocean wave heights. Int J Climatol 30(3):317–332

Uppala SM et al (2005) The ERA-40 re-analysis. Quart J Royal Meteorol Soc 131:2961–3012

IPCC (Intergovernmental Panel on Climate Change) (2000) Special report on emission scenarios. (Nakicenovic N, Nebojsa Nakicenovic, Alcamo J, Davis G, de Vries B, Fenhann J, Gaffin S, Gregory K, Griibler A, Jung TY, Kram T, La Rovere EL, Michaelis L, Mori S, Morita T, Pepper W, Pitcher H, Price L, Riahi K, Roehrl A, Rogner HH, Sankovski A, Schlesinger M, Shukla P, Smith S, Swart R, van Rooijen S, Victor N, DadiZ). Cambridge University Press, Cambridge, United Kingdom and New York

Chapter 4
Summary of Past and Future Climate Change

Summarizing the existing findings, the following conclusions can be drawn regarding the past changes in the wind climate:

- There has been an increase in storminess in the latter part of the twentieth century.
- However, the evidence suggests that at least in the NE Atlantic and in the northwest Europe the storminess was at the same level at the end of the nineteenth century as at the end of the twentieth century.
- There seems to have been a poleward shift of storm tracks over the last decades but fluctuations of similar magnitude have occurred earlier in the nineteenth and twentieth centuries.
- There is low confidence that reported long-term increases in tropical cyclone activity are robust, due to, inter alia, the fact that there have been changes in observing capabilities.
- There are many uncertainties in the historical tropical cyclone records and the understanding of the physical mechanisms linking tropical cyclone metrics to climate change is not complete.

The knowledge and confidence in projected *future* changes of storminess can be summarized as follows:

- There is some confidence that the frequency of the most intense storms will increase but not in all ocean basins.
- There are indications that tracks of extra-tropical cyclones will move poleward as a result of anthropogenic forcing.
- The extreme storminess seen over a hemisphere or for the north Atlantic and north Pacific in general show no significant signs of change in a warmer climate.
- There will be regional differences of the change in extreme storminess and wind speeds.
- Some regional studies indicate increased storminess over the northeast Atlantic and the North Sea regions but with different magnitude.
- There is low confidence in projected changes in location and tracks as well as duration, and areas of impact of tropical cyclones.

E. M. Bitner-Gregersen et al., *Ship and Offshore Structure Design in Climate Change Perspective*, SpringerBriefs in Climate Studies, DOI: 10.1007/978-3-642-34138-0_4, The Author(s) 2013

- The global frequency of tropical cyclones are likely to either decrease or remain essentially unchanged.
- The mean tropical cyclone maximum wind speed is likely to increase somewhat (+2 to +11 %) on a global scale but increases may not occur in all tropical regions.

For wave conditions the knowledge of *past* changes may be summarized as follows:

- There seems to have been an increase in significant wave height from the middle of the twentieth century to the early twenty-first century in the northern hemisphere winter in high latitudes in the north Atlantic and the north Pacific, with a decrease in more southerly latitudes of the northern hemisphere.
- If the record is extended back to late nineteenth century the picture changes, studies show that storminess and wave heights in late nineteenth/early twentieth century were about the same as near the end of the twentieth century.
- Even though some studies argue that the observed changes in wave conditions during the last part of the twentieth century are manifestations of climate change, it is still uncertain to what extent a climate change signal is detectable in the extreme wave heights.

It is uncertain to what extent future climate change will impact the extreme sea states that will be encountered by ocean going vessels. The reviewed studies show that:

- There will be regional increases in the sea states, more pronounced for extreme wind speed and SWH than for their means; e.g. the North and Norwegian Seas, immediately west of the British Isles, off the northwest of Africa, around 30°N from the east coast of the United states to 50°W and in the Pacific between 25 and 40°N and from the west coast of the United States to 170°W.
- The increases in extremes, represented by the 20-year return period of SWH or the highest storms in 20–30 years intervals are generally in the range 0.5–1.0 m in the North Atlantic, but larger increases can also be read off some graphs in the reviewed papers. Thus the increase may reach 10–18 % of the present 99th percentile H_s in the southern North Sea.

The differences between the projected changes in extreme sea states from studies using various global climate models, different climate forcing scenarios and a variety of downscaling approaches appear to be of the same order as the projected changes. Thus there are relatively large uncertainties associated with projected extreme under different climate, or emission scenarios. Our confidence in the projections is, therefore, limited.

Shortcomings of the reviewed publications include:

- Only a limited number of the factors that influence the projections has been studied simultaneously, e.g. global climate models versus climate forcing and statistical versus dynamical downscaling.

- The full effect of climate forcing scenarios has not been investigated, as emission scenarios with lowest and highest CO_2 emissions are not included in the reviewed literature.
- Focus has been on mean values and too low percentiles with only a few studies really considering extremes that are used in shipping, offshore and coastal design.
- The impact of choice of wave models has not been investigated sufficiently.
- The impact of choice of extreme value methods has not been investigated sufficiently.
- The reviewed studies have not been conducted with viewpoint of assisting the designer.

We recommend that international wave scientists and design engineers, offshore, coastal as well as ship, initiate a project that remedies these shortcomings.

Chapter 5
Potential Impact of Climate Change on Design of Ship and Offshore Structures

5.1 General

Structural failure of ship and offshore structures may result in loss of human life, severe environmental damage, and large economic consequences. Therefore ship and offshore structures must be designed with adequate safety and reliability, and their designs must be acceptable from an environmental and economic point of view. Environmental data and models represent a necessary and important input to load and response calculations of ship and offshore structures. They should be based on the state-of-the-art met-ocean description. Related relevant uncertainties in met-ocean data and models are also a part of such input.

To ensure that the designs are sufficiently safe and reliable, rules and offshore standards, including met-ocean, are developed by authorities or other competent organisations, such as e.g. Classification Societies. These rules and standards must then be adhered to by designers.

The design practice is moving gradually towards a more consistent probabilistic approach, for example: extremes are determined for a given return period (e.g. expected lifetime of the structure).

The previous chapters how wind and waves conditions may change in the twenty-first century. In the following chapters we show how the anthropogenic climate changes can be included in the current design practice. Particular focus is given to waves which have typically the largest impact on load and response calculations. Further, we demonstrate which consequences the observed and projected changes in wave climate may hav on tanker design.

5.2 Climate Change and Variability and Met-Ocean Design Criteria

Multi-decadal natural variability of climate due to the Earth's system dynamics, short term externally forced climate changes like volcanic activity and short term changes (10–12 years) in solar radiation to some extend have been taken care of in

E. M. Bitner-Gregersen et al., *Ship and Offshore Structure Design in Climate Change Perspective*, SpringerBriefs in Climate Studies, DOI: 10.1007/978-3-642-34138-0_5, The Author(s) 2013

design of ship and offshore structures by considering sufficiently long meteorological and oceanographic data records (typically much longer than 10 years). Note that the natural variability of the time scale larger than 50 years is usually not included in the data sets available for design.

Climate change due long term external forcing such as solar radiation and caused by changes in the Earth's orbit is neglected in a design process because of the large time scale of its occurrence.

It is, however, important to be aware that the natural climate variability can be of the same order of magnitude as the anthropogenic climate change and may mask it for several years to come. Further, the anthropogenic climate change is affecting the natural climate modes. Palmer (2008) suggests that change due to natural mode swap could be much larger than the direct anthropogenic change. Therefore the next 30–100 years' climate statistics may be affected significantly by it, a topic which is still not sufficiently investigated.

5.3 Risk-Based Approach Applied in Current Design

The traditional format of Classification Societies' Rules is mainly prescriptive, without any transparent link to an overall safety objective. IMO (1997, 2001, 2007) has developed Guidelines for use of the Formal Safety Assessment (FSA) methodology in rule development which will provide risk-based goal–oriented regulations. FSA consists of five inter-linked steps given in Table 5.1. When performing FSA for ship and offshore structures it is beneficial to apply Structural Reliability Analysis (SRA) in the risk assessment (step 2) and the cost-benefit assessment (step 4). Using this methodology, state-of-the-art met-ocean descriptions can be explicitly included in the rulemaking process.

The risk methodology based on the modern reliability methods is widely spread within the offshore sector.

In the risk based approach each event initiating structure failure may be represented by a limit state function that usually includes several causes responsible for its occurrence. These causes may be correlated or un-correlated.

Table 5.1 Steps of formal safety assessment (FSA)

Steps	In layman terminology	Professional language
1	*What might go wrong?*	Hazard identification
2a	*How often or how likely?*	Frequencies or probabilities
2b	*How bad?*	Consequences
2c	*How to model?*	Risk = Probability consequence
3	*Can matters be improved?*	Identify risk management options
4	*What would it cost and how much better would it be?*	Cost benefit evaluation
5	*What actions are worthwhile to take?*	Recommendation
IMO	What action to take?	Decision

The basic problem in Structural Reliability Analysis (SRA), see e.g. Madsen et al. (1986), Skjong et al. (1995), Ditlevsen and Madsen (1996) may be formulated as the problem of calculating the (small) probability that

$$g(X_1, X_2, \ldots X_N) < 0 \qquad (5.1)$$

where $X = (X_1, X_2, \ldots X_N)$ is a vector of basic random variables and $g(X)$ is referred to as the limit state function that describes the failure set, the failure surface, and the safe set, i.e.

$$g(\mathbf{X}) \begin{cases} > 0 & \text{for } \mathbf{X} \text{ in safe set} \\ = 0 & \text{for } \mathbf{X} \text{ on the limit state surface} \\ < 0 & \text{for } \mathbf{X} \text{ in failure set} \end{cases} \qquad (5.2)$$

$g(X)$ is a random variable and its distribution is determined by the g-function and the probabilistic model for the basic variables. The variables describe functional relationships in the physical model and the randomness of parameters in the model. A parameter of a variable may be a function of coordinates of other variables so that a network structure for dependencies between variables can be defined. Statistical dependence between the variables can also be modelled through correlations.

Equation (5.1) describes the physical problem while the random variables and distributions describing them are defining a probabilistic model.

An event, $E(X)$, is a subset of the sample space for the stochastic process i.e., a subset of all the possible outcomes of the stochastic process. An event may be defined through a functional relationship

$$E(\mathbf{X}) = \{\mathbf{x}; g(\mathbf{x}) \leq 0\} \qquad (5.3)$$

The event identifies the outcomes of interest while the random variables X define the nature of a stochastic process.

The event probability, P_E, is the probability that an outcome of the stochastic process X yields the event E,

$$P_E = P(E(X)) \qquad (5.4)$$

In Structural Reliability Analysis the reliability index β_R is defined to be the argument of the standard normal distribution (Φ) which yields one minus the event probability, and used as a measure i.e.

$$\beta_R = \Phi^{-1}(1 - P_E) = -\Phi^{-1}(P_E) \qquad (5.5)$$

where P_E is the failure probability.

This basic problem may be transformed into an equivalent problem where the stochastic variables are transformed into a standard-normal-space, i.e. the space of decorrelated normally distributed variables with zero mean and unit standard deviation. This transformation (Rosenblatt 1952) is

$$\left[\begin{array}{c} u_1 = \Phi^{-1}(F(x_1)) \\ \ldots \\ u_i = \Phi^{-1}(F(x_i|x_1, x_2, \ldots, x_{i-1})) \\ \ldots \ldots \\ u_n = \Phi^{-1}(F(x_n|x_1, x_2, \ldots, x_{n-1})) \end{array}\right. \tag{5.6}$$

where $x = (x_1, x_2, \ldots, x_j \ldots, x_n)$ is the basic vector and $\mathbf{u} = (u_1, u_2, \ldots, u_j, \ldots u_n)$ is the transformed vector, F denotes the cumulative probability function and Φ is the standard-normal distribution.

The reliability methodology allows for consistent treatment of uncertainties and provides probabilities where uncertainties can be included. By adopting SRA sensitivity studies can be carried out and importance of analysed parameters to the failure probability identified. These results may be used in various ways, for example to present risk estimates with and without epistemic uncertainties included.

The sensitivity of a calculated probability (or of the first statistical moments of the g-variable) to changes in one or more parameters in the model can be calculated by use of the *Parametric Sensitivity Factor* defined as the derivative of the reliability measure (probability or reliability index or statistical moments) with respect to a parameter, say ϑ, Hohenbichler and Rackwitz (1986). The change in the failure probability given by a change in a parameter ϑ is estimated as

$$P_E(\vartheta + \Delta\vartheta) = P_E(\vartheta) + \frac{\partial P_E(\vartheta)}{\partial\vartheta}\Delta\vartheta \tag{5.7}$$

The parameter ϑ may be a fixed valued variable or a parameter in a function or a distribution (e.g. a mean value or a standard deviation in a distribution).

The *Uncertainty Importance Factor*, defined as the square of the *Geometrical Sensitivity Factor*, is another parameter which the reliability methods provide. The *Geometrical Sensitivity Factor*, α_μ, is defined to be the derivative of the reliability index β_R with respect to the mean μ of the corresponding u-space variable, Hohenbichler and Rackwitz (1986). Thus

$$\alpha_\mu = \frac{\partial\beta_R}{\partial\mu} \tag{5.8}$$

The *Uncertainty Importance Factor* indicates the importance of modelling the random variable X as a distributed variable rather than as a fixed valued variable, the median of the distribution being the fixed value. In other words, the *Uncertainty Importance Factor* of the ith variable roughly gives the fraction of the total uncertainty which is caused by uncertainty in this variable.

Thus the reliability methods allow quantifying in a probabilistic way the uncertainties in the different parameters that govern the structural integrity. This allows reliability assessment of structural components or a structure. Further reliability-based design of a structural component (or a structure) provides a means to satisfy target reliability with respect to specific modes of failure. The probabilistic approach can be used for calibration of partial safety factors in the development of Load and Resistance Factor Design (LRFD) codes (see Ronold and Skjong 2002), and for development of acceptance criteria for structural designs, confer DNV (1992), ISO 2394 (1998), Skjong et al. (1995), Bitner-Gregersen et al. (2002), Skjong and Bitner-Gregersen (2002), Hørte et al. (2007a, b).

Standard software allowing for carrying out structural reliability calculations has been available within the industry since the mid-eighties. Also, complicated non-linear effects can be included by embedding a time domain simulation code in a reliability code, like the probabilistic analysis code PROBAN® (Det Norske Veritas 2002). The program includes the First Order Reliability Method, FORM and the Second Order Reliability Method, SORM which have been introduced to solve the high reliability problems often encountered in the studies of structural safety. These methods are theoretically justified by asymptotic theory and attractive in many applications. For more likely events a number of sampling methods are made also available like Monte Carlo, directional, axis-orthogonal and Latin-Hypercube simulation.

When discussing impact of extreme waves, and met-ocean conditions generally, on marine structures a distinction needs to be made between ship structures and offshore structures. Even though the same basic principles prevail for hydrodynamic loads on ships and offshore structures, actual problems and methods for assessing these loads in the design stage are quite different. Further, different wave data and to some extent different wave models are used for defining design and operational conditions for these two types of structures.

Sailing (non-stationary) ships do not include vessels that operate at a fixed location (e.g. FPSO's). A salient feature of ship hydrodynamics is the non-zero forward speed. Further, as ships are sailing they are exposed to varying wave environment. This fact needs to be taken into consideration when specifying design and operational criteria.

Unlike ship structures, offshore structures normally operate at fixed locations and often represent a unique design. Therefore site specific environmental data are usually required.

To have a clear and consistent approach for determining design loads, we need to define the limit state categories and the scenarios we design for. In the offshore industry the following well proven terminology (e.g. ISO) is applied which is starting to be accepted also within the shipping industry (see e.g. DNV 1992):

Ultimate Limit State (ULS) corresponding to the maximum load carrying resistance.

Fatigue Limit State (FLS) corresponding to the possibility of failure due to the effect of cyclic loading.

Serviceability Limit State (SLS) corresponding to the criteria applicable to normal use or durability.

Accidental Limit State (ALS) corresponding to the ability of the structure to resist accidental loads and to maintain integrity and performance due to local damage or flooding.

Climate changes of met-ocean conditions will impact all limit state categories but it is expected that ULS and ALS will be most affected.

In the design process, international standards are followed to calculate ship structural strength and ship stability during extreme events. The return period for ship structures is 20 years (ULS). Recently, IMO has increased the design life to 25 years in Goald-Based Standards (GBS) for bulk and tankers with length over 150 m. Accidental Limit State, ALS, (corresponding to the ability of the structure to resist accidental loads and to maintain integrity and performance due to local damage or flooding) checks cover grounding, collision, and fire and explosion. An extreme weather event check is not included in ALS, as explained in Bitner-Gregersen et al. (2003), Hørte et al. (2007a).

Offshore structures (including FPSOs) follow a different approach to design of ship structures and are designed for the 100-year return period (ULS). The Norwegian offshore standards (NORSOK Standard (2007)) requires that there must be enough room for the wave crest to pass beneath the deck to ensure that a 10000-year wave load does not endanger the structure integrity (ALS).

5.4 Risk-Based Approach Including Climate Changes

Design of ship and offshore structures will be affected by changes of surface ocean temperature, wind, waves, sea water level and ice reported by IPCC (2007, 2011, 2012) although sensitivity to the climate changes may vary for different structure types. Attention also has to be given to marine growth on ship and offshore structures, which is expected to increase significantly due to global warming. This development may, however, be compensated by better coating.

Three aspects of met-ocean description in particular need to be considered when discussing possible impact of climate change on design and operations of ship and offshore structures (Bitner-Gregersen and Eide 2010):

- Long-term variations (anthropogenic changes and natural variability) of climate
- Extreme weather events
- Uncertainty modelling

Long-term variations (several decades' variations) of meteorological and oceanic conditions and their statistical characteristics will affect the currently used met-ocean data bases and, therefore, the design and operation criteria derived for ship and offshore structures. Predicted load and response projections will also be affected. Changes in extreme weather events may impact long-term statistical description of met-ocean environment as well as the current methodology and calculation procedures for prediction of short-term variations (20-min up to 3–6 h) of ship and offshore structures' loads and responses, and will need to be accounted for. Note that information about long-term and short-term variations of met-ocean conditions is combined in a design process. Specification of uncertainties of climate change projections is essential and will decide accuracy of met-ocean design and operational criteria provided.

It is also important to be aware that changes, like increase in storm activity (intensity, duration and wind fetch) in some regions (still low confidence in these projections, IPPC 2012), may lead to secondary effects such as increased frequency of occurrence of extreme wave events. More intense swell might also be expected. The frequency of occurrence of combined wave systems like wind sea and swell (one, or several swell components) may increase in some ocean areas due to increase of storm intensity and change of storm tracks. Combination of wind sea and swell may consequently lead to more frequent extreme events (Onorato et al. 2006; Shukla et al. 2006; Toffoli et al. 2011b), something not investigated sufficiently. Vulnerability to hurricane storm-surge flooding may increase if the projected rise in sea level due to global warming occurs. These extreme weather events will affect long-term met-ocean statistics and may have impact on current methodology and procedures for load and response calculations.

Climate changes of met-ocean conditions and relevant uncertainties will need to be an integrated part of the risk-based approach as illustrated schematically in Fig. 5.1. Possible inclusion of rogue waves in the risk based approach is discussed by Bitner-Gregersen et al. (2003, 2008). Wave directional spreading will represent an important characteristic of rogue waves needed to be taken into consideration, e.g., Toffoli and Bitner-Gregersen (2011a).

Identification of uncertainties and their quantification represents important information for risk assessment of ship and offshore structures (see Fig. 5.1). How to handle uncertainties in a risk based rule and offshore standard format is well

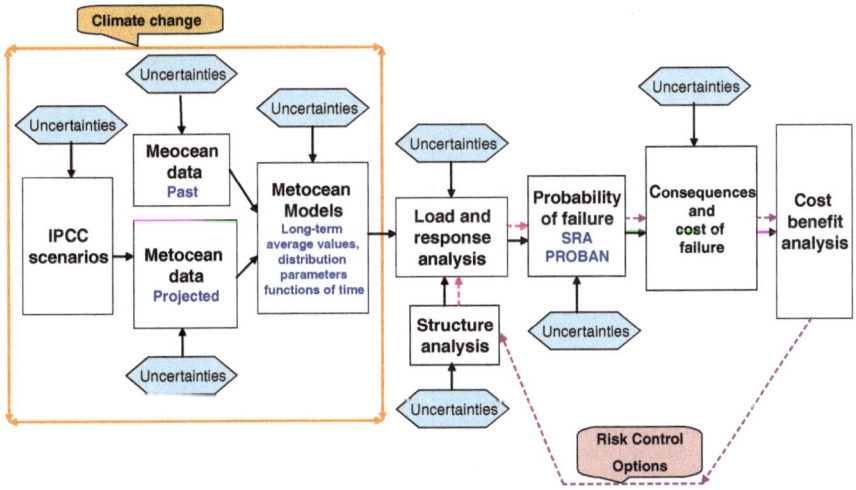

Fig. 5.1 Risk based approach, overview of interfaces and climate change is integrated

established (Madsen et al. 1986), see also e.g. Bitner-Gregersen et al. 2002, and Hørte et al. 2007a, b). The significance of uncertainty modelling of met-ocean conditions will increase when climate change is considered as no field observations will be available for validation of the projected future climate. Uncertainties discussed in Chap. 3 will need to be considered.

Climate change trends have non-stationary character which is not accounted for in current design practice of ship and offshore structures. To be able to design for climate change time-dependent statistical descriptions need to be adopted. Statistical extreme value analysis, as currently used in the met-ocean community, has to be upgraded to take into account the non-stationary character of current climate, in terms of both climate change trends and natural variability cycles. Spatial–temporal models, like the Bayesian hierarchical space–time model e.g., Vanem et al. (2011a, b; Vanem 2012), are expected to play an important role in this development. Climate trends obtained from the climate/wave models as well as Bayesian hierarchical space–time models will need to be incorporated into joint met-ocean description currently used in design. These non-stationary climate change trends will be a part of the risk-based approach as illustrated schematically in Fig. 5.1.

A distinction will need to be made between existing structures and new ones when evaluating impact of climate change on ship and offshore structures design. SRA is recommended to be used for checking whether the existing structures will maintain the same safety level as current design. It is too early to conclude which revisions will need to be introduced in the current ship and offshore structure design and what economic consequences they will have. These considerations will need to be based on cost benefit analysis as illustrated in Skjong and Bitner-Gregersen (2002).

In the following section we demonstrate what impact climate changes may have on current design practice of tankers. The results presented are based on the state-of-the-art knowledge on climate change projections and therefore some revision may be required when new investigations regarding climate change arrive in the future.

References

Bitner-Gregersen EM, Eide LI (2010). Climate change and effect on marine structure design. DNVRI Position Paper No.1 http://www.dnv.com/resources/position_papers/new_frontiers.asp
Bitner-Gregersen EM, Hovem L, Skjong R (2002) Implicit reliability of ship structures. In: Proceedings of OMAE 2002, Oslo, 23–28 June 2002
Bitner-Gregersen EM, Hovem L, Hørte H (2003) Impact of freak waves on ship design practice. In: Proceedings of maxwave final meeting, Geneva, 8–10 Oct 2003

Bitner-Gregersen EM, Toffoli A, Onorato M, Monbaliu J (2008) Implications of nonlinear waves for marine safety. In: Proceedings of rogue waves 2008 workshop, Brest, France, 13–15 Oct 2008, http://www.ifremer.fr/web-com/stw2008/rw/

Ditlevsen O, Madsen HO (1996) Structural reliability methods. Wiley, Chichester, West Sussex

DNV (1992) Classification Note 30.6: structural reliability analysis of marine structures, July 1992

DNV (2002) PROBAN theory, general purpose probabilistic analysis program, the author L. Tvedt, Version 4.4, Høvik, Norway

Hohenbichler M, Rackwitz R (1986) Sensitivity and importance measures in structure relaibility. Civ Eng Syst 3:203–209

Hørte T, Skjong R, Friis-Hansen P, Teixeira AP, Viejo de Francisco F (2007a) Probabilistic methods applied to structural design and rule development. In: Proceedings of the RINA conference "development of classification and international regulations", London, 24–25 Jan 2007

Hørte T, Wang G, White N (2007b) Calibration of the hull girder ultimate capacity criterion for double hull tankers. In: Proceedings of 10th international symposium on practical design of ships and other floating structures, Houston

IMO (1997) Interim guidelines for the application of formal safety assessment (FSA) to the IMO rule making process. MSC Circ.829/MEPC/Circ.335

IMO (2001). Guidelines for formal safety assessment for the IMO rule making process. MSC/ Circ.1023—MEPC/Circ.392

IMO (2007) Formal safety assessment, consolidated text of the guidelines for formal safety assessment (FSA), for use in the IMO rule-making process (MSC/Circ.1023—MEPC/ Circ.392), MSC83/INF.2

IPCC (2007) Climate Change (2007) The physical science basis. In: Solomon S, Qin D, Manning M, Chen Z, Marquis M, Averyt KB, Tignor M, Miller HL (eds) Contribution of working Group I to the fourth assessment report of the intergovernmental panel on climate change. Cambridge University Press, Cambridge, 996 pp

IPCC (2011) Summary for policymakers. In: Field CB, Barros V, Stocker TF, Qin D, Dokken D, Ebi KL, Mastrandrea MD, Mach KJ, Plattner G-K, Allen S, Tignor M, Midgley PM (eds) Intergovernmental panel on climate change special report on managing the risks of extreme events and disasters to advance climate change adaptation. Cambridge University Press, Cambridge

IPCC (2012) Managing the risks of extreme events and disasters to advance climate change adaptation. In: Field CB, Barros V, Stocker TF, Qin D, Dokken DJ, Ebi KL, Mastrandrea MD, Mach KJ, Plattner G-K, Allen SK, Tignor M, Midgley PM (eds) A special report of working Groups I and II of the intergovernmental panel on climate change. Cambridge University Press, Cambridge, 582 pp

ISO 2394 (1998) General principles on reliability for structures. ISO TC 98/SC 2/ WG 1, 2nd edn 1998-06-01

Madsen HO, Krenk S, Lind NC (1986) Methods of structural safety. Prentice-Hall, Enlewood Cliffs, NJ 07632

NORSOK (2007) Standard N-003: action and action effects, Rev. 2. http://www.standard.no/ pronorm-3/data/f/0/03/78/7_10704_0/N-003d2r2.pdf

Onorato M, Osborne AR, Serio M (2006) Modulation instability in crossing sea states: a possible mechanism for the formation of freak waves. J Phys Rev Lett 96:014503

Palmer T (2008) Introduction to CLIVAR exchanges. http://www.clivar.org

Ronold K, Skjong R (2002) The probabilistic code optimisation module PROCODE. In: Joint committee on structural safety (JCSS) workshop on reliability based code calibration, Zurich, 21–22 March 2002

Rosenblatt M (1952) Remarks on a multivariate transformation. Ann Math Stat 23:470–472

Shukla PK, Kaurakis I, Eliasson B, Marklund M, Stenflo L (2006) Instability and evolution of nonlinearly interacting water waves. J Phys Rev Lett 97:094501

Skjong R, Bitner-Gregersen EM (2002) Cost effectiveness of hull girder safety. In: Proceedings of OMAE-2002-28494, Oslo

Skjong R, Bitner-Gregersen EM, Cramer E, Croker A, Hagen Ø, Korneliussen G, Lacasse S, Lotsberg I, Nadim F, Ronold KO (1995) Guidelines for offshore structural reliability analysis—general. DNV Report No 95–2018

Toffoli A, Bitner-Gregersen EM (2011) Extreme and rogue waves in directional wave field. Open Ocean Eng J 4:24–33

Toffoli A, Bitner-Gregersen EM, Osborne AR, Serio M, Monbaliu J, Onorato M (2011b) Extreme waves in random crossing seas: laboratory experiments and numerical simulations. Geophys Res Lett 38:L06605, 5. doi: 10.1029/2011

Vanem E (2011) Long-term time-dependent stochastic modelling of extreme waves. Stoch Environ Res Risk Assess 25(2):185–209

Vanem E (2012) A stochastic model for long-term trends in significant wave height with a CO2 regression component. In: Proceedings of OMAE 2012 conference, Rio de Janeiro, 1–6 July 2012

Vanem E, Huseby AB, Natvig B (2011b) A Bayesian hierarchical spatio-temporal model for significant wave height in the North Atlantic. Stoch Environ Res Risk Assess 1–24 doi:10.1007/s00477-011-0522-4

Chapter 6
Consequences of Wave Climate Change for Tanker Design

6.1 Current Design Wave Database for Ship Structures

The need to improve the availability, quality, and reliability of environmental databases (mainly wind and wave data for ships) has been identified by various international professional organisations (e.g. ISSC 2009) as well as Classification Societies. Several studies have attempted to quantify the uncertainties due to insufficient knowledge of the wave climate (e.g. Bitner-Gregersen and Guedes Soares 2007) resulting in differences in long-term ship responses of up to 150 % of a nominal value, e.g. Guedes Soares and Trovao (1991), Bitner-Gregersen et al. (1995a). This high uncertainty may lead to overdesign/underdesign of ships, with consequent significant economic/risk impact.

Visual observations of waves collected from ships in normal service are currently used in the design of ship structures. Hogben and Lumb (1967) data were originally applied, but these were later replaced by the more reliable Global Wave Statistics (GWS) visual observations (British Maritime Technology, BMT 1986). In the GWS atlas the ocean is divided into 104 regions as shown in Fig. 6.1. The visual data represent a sufficiently long observation history to provide reliable global climatic statistics. Wind speeds (Beaufort Scale) and directions, and wave heights in a coarse code have been reported since 1854. Observations of wave height, period, and direction have been collected from ships in normal service all over the world since 1949, and are made in accordance with guidance notes from the World Meteorological Organisation (2001, 2003).

The utility of visual observations depends on appropriate calibration versus accurate measurements of the wave characteristics. BMT (1986) compared the GWS marginal distributions for wave heights and wave periods with instrumental Shipborne Wave Recorder and National Oceanic and Atmospheric Administration (NOAA) buoy data for different locations and concluded that the wave heights and periods for which statistics were given corresponded to measured values. However, the accuracy of the GWS data has been questioned, especially concerning the wave period (e.g., Bitner-Gregersen and Cramer 1994; Bitner-Gregersen et al. 1995a, b).

E. M. Bitner-Gregersen et al., *Ship and Offshore Structure Design in Climate Change Perspective*, SpringerBriefs in Climate Studies, DOI: 10.1007/978-3-642-34138-0_6, The Author(s) 2013

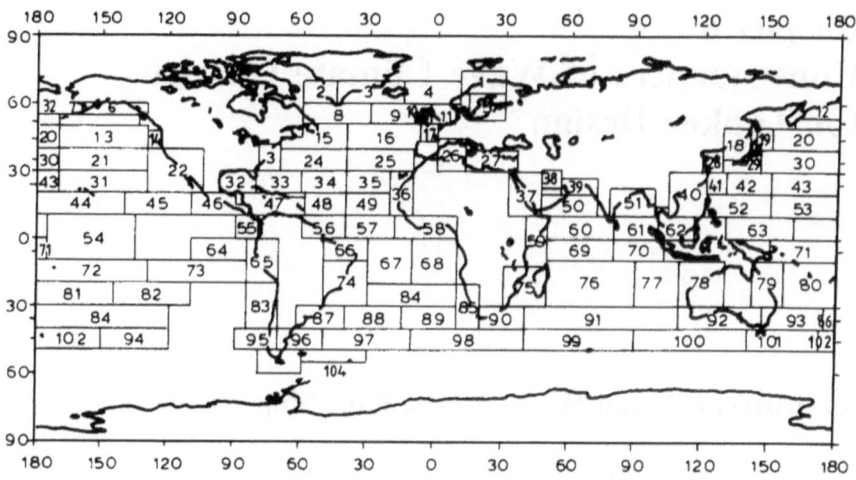

Fig. 6.1 Global wave statistics zone designation, BMT (1986)

Several studies have addressed uncertainties in the Global Wave Statistics data and their effects on ship loads and responses, as well as on fatigue damage (Chen and Thayamballi 1991; Bitner-Gregersen et al. 1993, 1995a), and have concluded that the GWS should be used with care. In 1995, a general procedure for correction of the mean wave period identified bias of the GWS data was suggested by Bitner-Gregersen et al. (1995b), and a scatter diagram for the North Atlantic, representing an average diagram for ocean zones 8, 9, and 15, was proposed for ship design. The scatter diagram was numerically generated from a joint distribution of wave height and wave period fitted to the GWS data, where the wave period distribution was corrected by the period bias. The original GWS data were not used. This was not only because of the wave period inaccuracy, but also, as in the GWS atlas, the last wave height class is in the range 11–12 m, indicating that observations of wave height beyond 12 m were summed up by BMT (1986) in this class. The numerically simulated North Atlantic scatter diagram altered the last wave height class to a height class range 16–17 m.

This adjusted scatter diagram has been adopted as a DNV standard in 2000 (today DNV RP-C205 2010). Later, on a request from the International Association of the Classification Societies (IACS), the North Atlantic scatter diagram was extended by DNV to include additionally the ocean area 16 (see Fig. 6.1), using the procedure described above. The extended scatter diagram is included in IACS Recommendations No. 34 (2000), as well as in the DNV Recommended Practice (2010), and is regarded as a 20-year return period scatter diagram for ship design. It should be noted that this updated scatter diagram is slightly more conservative than the exact 20-year scatter diagram would be. For ship load and response calculations, a joint distribution of significant wave height and zero-crossing wave period is usually fitted to the data given by the scatter diagram, and the 20-year (or 25-year) return period is then derived from the fitted model. The DNV (2010)

Recommended Practice is used for both ship and offshore structures. Note that the North Atlantic scatter diagram is also used for design of Floating Production Storage and Offloading systems (FPSOs), while other offshore structures are designed for location-specific met-ocean environments.

The necessity of replacing the current wave database for ship design by measured data, or by a combination of numerical and measured data, has increasingly become a subject for discussion within DNV and IACS in recent years. Currently, two other sources of global met-ocean climate are available, in addition to the ship observations. These are data from numerical wave prediction models and satellite data. Based on these data and on instrumental observations, several global databases have been developed. However, predictions based on these databases show significant discrepancies (see Bitner-Gregersen and Guedes Soares 2007), and therefore they are presently unsuitable for establishing new design wave data statistics for ships. This topic needs to be revisited continuously as wave databases are under development.

6.2 Hull Girder Collapse in Extreme Sagging Conditions

In light of the findings summed up in Chap. 4 the next sections will illustrate the potential impacts that climate change may have on the design by using tanker design as an example. More specifically, the impact on hull girder collapse of tankers has been studied. The IACS Common Structural Rules for Tankers, IACS (2010), has been used to demonstrate this effect on structure design.

During the development of the IACS Common Structural Rules for Tankers (CSR) Structural Reliability Analysis (SRA) was used as a tool to calibrate a new hull girder ultimate strength criterion. This rule criterion was introduced as an explicit control of the most critical structural failure mode identified as sagging failure of a loaded tanker in severe weather conditions.

6.2.1 Set-Up of the Structural Reliability Analysis (SRA)

When establishing the IACS Common Structural Rules for Tankers, IACS (2010), a probabilistic model for the midship vertical bending moment, due to still water and wave loads, as well as for the ultimate bending moment capacity was developed and applied. A test set of five ships, ranging from Product tanker to VLCC, were analysed. Here, these cases have been further investigated by SRA to demonstrate the impact of expected wave climate change on the hull girder failure probability of ship structure design.

The methodology used by the IACS Common Structural Rules for Tankers adopted in the present study is described in Bitner-Gregersen et al. (2002) and Hørte et al. (2007a, b). The probabilistic analysis program PROBAN (DNV 2002) is applied for the calculations. Five oil tankers, ranging from Product tanker to VLCC are considered. Note that the ship length is ranging from 174.5 to 320 m. The ship dimensions are given in Table 6.1. For detail description of input used in the study the reference is made to Hørte et al. (2007b).

Ship structural strength and ship stability are calculated, following international standards, in extreme events with an occurrence of once in every 20 years [Ultimate Limit State (ULS)].

The projections of wave climate due to climate change in the North Atlantic, discussed in Sect. 2.6 show an increase of extreme significant wave height from ca. 0.5–1.0 m with the uncertainty of the same size. Note that these numbers refer to the end of the twenty-first century.

The results presented herein are limited to the structural collapse of ships owing to buckling of ship decks in extreme sagging conditions (see Fig. 6.2); Bitner-Gregersen et al. (2011), Bitner-Gregersen and Skjong (2011). The potential consequence is total loss of ship and crew. For a tanker in full loading conditions, the assumption of structural collapse owing to ship deck buckling in the extreme sagging condition seems to be rather close to realistic collapse mode as the contributions to the moment capacity from longitudinal bulkheads and ship sides are all small.

The Ultimate Limit State (ULS) failure criterion is expressed in terms of a limit state function $g(X)$ [see Eqs. (5.1–5.2)] that describes the failure set (no hull girder collapse), the failure surface, and the safe set (hull girder collapse).

Thus, the probability of failure due to buckling of ship deck is (see Fig. 6.2).

$$P_f = P(g(X_1, X_2, \ldots X_N) \leq 0) \tag{6.1}$$

with the corresponding reliability index β_f defined as

$$\beta_f = -\Phi^{-1}(P_f) \tag{6.2}$$

where Φ denotes the standardized cumulative normal distribution function.

For hull girder collapse the following limit state function is used

$$g(\mathbf{X}) = M_U \cdot X_R - (M_{WV} \cdot X_{st} \cdot X_{nl} + M_{SW} \cdot X_{SW}) \tag{6.3}$$

where M_{sw} is the random still water bending moment, M_{wv} is the random wave bending moment, M_u is the random ultimate capacity, and X_R, X_{st}, & X_{nl} and X_{sw} represent model uncertainty factors for the capacity, wave-induced bending

Table 6.1 Test ships

Case	Ship type	Lpp (m)	Breadth (m)	Depth (m)
1	Suezmax	263	48	22.4
2	Product	174.5	27.4	17.6
3	VLCC 1	320	58	31
4	VLCC 2	316	60	29.7
5	Aframax	234	42	21

Fig. 6.2 Hull girder collapse in extreme sagging conditions

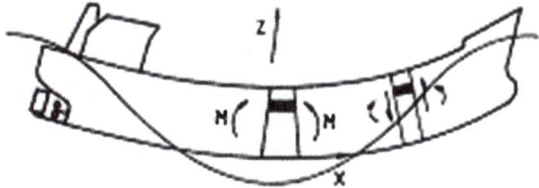

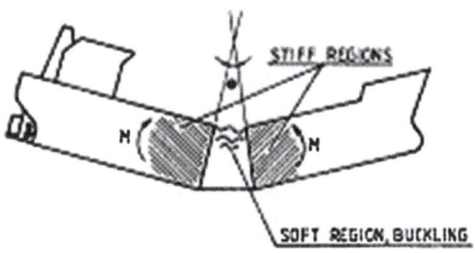

moments and still water bending moment, respectively. The set-up of the analysis is illustrated in Fig. 6.3. The load analysis is carried out prior to the structural reliability analysis, with transfer functions stored on an interface file. This enable short- and long-term response calculation within the probabilistic analysis. The capacity calculation is an integrated part of the probabilistic analysis. In addition to the probability of failure, the SRA provides uncertainty importance factors, sensitivity factors and the design point which represents the most likely values of the variables at failure (see Sect. 5.2). The first order reliability analysis method (FORM) has been used.

Fig. 6.3 The set-up of the analysis

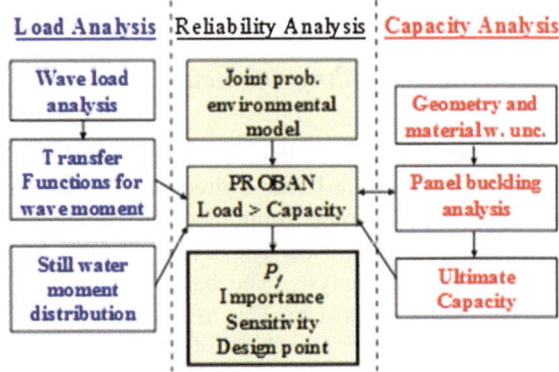

6.2.2 Still Water Bending Moment

In general for tankers, ballast conditions induce hogging moments whereas loaded conditions induce sagging moments. Therefore, sagging failure in ballast condition is rather unlikely. In addition, the wave moment is generally somewhat larger in loaded than in ballast conditions.

The capacity in hogging is usually significantly higher than in sagging due to the double bottom. Sagging failure is governed by the ultimate capacity of the deck and is considered in the present work. The consequences of sagging failure are also more critical than those of hogging failure, both from the economic and environmental points of view, since the ship is fully loaded.

The following steps have been made for the assessment of the probability distribution of the random still water bending moment (SWBM), M_{sw}:

1. Identification of all seagoing loaded conditions in the loading manual, see also Gran (1992).
2. Removing emergency ballast and segregated/transitory/group load conditions (that often give hogging).
3. Calculation of the mean value and the standard deviation of the identified loading conditions, assuming equal weighting for each condition.

Based on this, a generic distribution of the midship still water bending moment for the loaded (sagging) condition has been established. A normal distribution has been fitted to the data, with a mean value of 0.7 and a standard deviation 0.2 times the maximum value from any full-load condition listed in the loading manual. There is no upper threshold applied to the distribution. In other words, there is a chance that the SWBM may attain a value that exceeds the maximum value in the loading manual, at a probability of 7 %.

Furthermore a model uncertainty factor, X_{sw}, has been defined by a normal distribution with a mean value of 1.0 and a standard deviation of 0.1 [N(1.0, 0.1)], and multiplied to the distribution of the still water bending moment. This uncertainty was included to reflect the uncertainty between the actual SWBM and the corresponding calculated value in the loading manual.

6.2.3 Wave Bending Moment

M_{wv} is the random wave bending moment (WVBM). The structural response due to waves is based on linear hydrodynamic analysis. Results in terms of transfer functions (or Response Amplitude Operators, RAOs) for the midship vertical bending moment are used. The short- and long-term response is computed within the probabilistic analysis. The basic assumption is a narrow banded Gaussian response in each sea state. This assumption implies Rayleigh distributed maxima for a given sea state, for which a Gumbel type extreme value distribution can be derived. Finally, the annual extreme value distribution is obtained assuming independence between sea states. The approach using transfer functions with PROBAN is documented by Mathisen and Birknes (2003). This model captures the uncertainty in the short term response.

The annual probability of failure is calculated taking into account the relevant fraction of the year for which the ship is in the fully loaded condition and at sea; assumed to be 42.5 % of the year.

In heavy weather at sea, the ship is most likely to operate in head seas, or nearly head seas. The waves tend to be more long-crested in extreme sea states than in lower sea states. With these considerations, the analyses are carried out with the assumption of wave spreading corresponding to a $\cos^4$ directional spreading function, and a triangular distribution of main heading with limits ± 30 around head sea.

The Pierson Moskowitz (PM) wave spectrum is applied in the analyses. Furthermore, an assumption of zero-speed is adopted. This assumption is satisfactory, because in an ULS condition the forward speed is very low (less than 5 knots), and the load analysis is not sensitive to small forward speed variations as shown e.g. by Bitner-Gregersen et al. (2002).

A model uncertainty for the response calculation is applied in terms of a normally distributed uncertainty factor with a mean value of 1.0 and a coefficient of variation of 0.1. This uncertainty factor is assumed to cover uncertainty in the linear results, including the effect of uncertainty in the wave spectrum. Reference is made to DNV (1992).

Furthermore, the use of a linear analysis for the bending moment response in extreme weather is a simplification. The problems are inherently non-linear dealing with large-amplitude non-linear wave fields and the variable geometry of the ship's hull as it comes in and out of the water as well as with slamming, wave breaking and green water on deck, ISSC (2000a). It is difficult to conclude on a "correct" model uncertainty to account for non-linear effects; e.g. the extent of green water on deck will tend to reduce the sagging moment. Bottom slamming and whipping may lead to an increase in the sagging moment, but is not so likely to occur in loaded conditions.

Hence, model uncertainty factors X_{st} for the linear response calculation and X_{nl} for the nonlinear effects are applied, both N(1.0, 0.1).

A joint environmental model, with a 3-parameter Weibull distribution for the significant wave height H_s and a conditional lognormal distribution for the zero-crossing wave period, T_z due to Bitner-Gregersen (1988) is applied (see also Bitner-Gregersen and Haver 1989, 1991; Mathisen and Bitner-Gregersen 1990)

$$F_{Hs}(h_s) = 1 - \exp\left(-\left(\frac{h_s - \gamma}{\alpha}\right)^{\beta}\right) \qquad (6.4)$$

$$F_{Tz|Hs}(t_z|h_s) = \Phi\left(\frac{\ln t_z - \mu_{\ln T_z}}{\sigma_{\ln T_z}}\right) \qquad (6.5)$$

where

$$\mu_{\ln T_z} = a_0 + a_1 \cdot h_s^{a_2} \tag{6.6}$$

$$\sigma_{\ln T_z} = b_0 + b_1 \cdot \exp(b_2 \cdot h_s) \tag{6.7}$$

The parameters α, β, γ, a_i and b_i are site specific, and all depend on the sailing route of the ship over the design life. Note that the model can also be applied to the spectral peak period T_p.

The model described by Eqs. (6.4–6.7) has been fitted to the North Atlantic scatter IACS diagram (IACS 2000) which represents visual observations with correction of the wave period due to Bitner-Gregersen et al. (1995a, b) see Sect. 5.5.1.

The environmental model is modified to reflect the climate change by shifting the distribution of the significant wave height by a constant value corresponding to the specified increase. The formulation of the conditional distribution of the zero-crossing period is kept unchanged. This simplification is considered to be acceptable for extremes but not for fatigue calculations. This idea is further developed in Vanem and Bitner-Gregersen (2012) who provide closed form expressions for the modified Weibull parameters including climate trends expressed in terms of the mean value and the standard deviation.

6.2.4 Combination of Still Water and Wave Bending Moment

The still-water bending moment is added to the wave moment by linear super-position. Two different combinations, following Turkstra's combination Rule, (Turkstra 1970), are evaluated:

(a) An annual extreme value of the wave induced moment together with a random value of the still water moment.
(b) An annual extreme value of the still water moment together with an extreme value of the wave moment during one voyage.

Depending on the relative magnitude between the two contributions, the variability and the duration of a voyage, either of these combinations may be governing. In the present tanker study it appears that the extreme wave load is most critical and combination (a) is therefore governing for the probability of failure. This is also found in other studies; e.g. Bach-Gansmo and Lotsberg (1989) and Kaminski (1997).

6.2.5 Ultimate Bending Moment Capacity

M_u is the random ultimate capacity, and it is calculated according to the single step method given in the IACS Common Structural Rules for tankers. The panel buckling program PULS (DNV 2009) is used to compute the ultimate plate field buckling capacity, here as an integrated part of the structural reliability analysis.

The model applied to describe the ultimate moment capacity M_u of the hull girder, accounting for mode interaction effects between local and overall Euler buckling modes, is based on a non-linear buckling model. For a more detailed description of this capacity model, see Steen (1996).

In this model, the moment capacity is defined as the product of the modified sectional modulus W_u for the deck and the ultimate strength of the deck panels, i.e.

$$M_u = W_u \cdot \sigma_u \tag{6.8}$$

W_u and σ_u are both stochastic variables and depend on the buckling characteristics of the deck structure.

The stiffened plate model for the deck plating is assumed to have the same proportions and characteristics over the entire deck area. The deck buckling strength is a function of the geometrical dimensions of the plate and stiffeners, the out-of-straightness of the plate and stiffeners, the Young's modulus, and the yield stress of the material.

Uncertainty with respect to the yield strength of the material is accounted for in the analysis. The distribution of the yield strength is derived from its characteristic value which represents the lower 5 % fractile. A coefficient of variation of 0.08 is used for mild steel (for the Product tanker) and 0.06 is used for high strength steel (for the remaining test ships). These values are taken from the DNV (1992) and Skjong et. al (1995), and are commonly applied. In ISSC (2000b) comparable yield strength coefficients of variation of 0.09 and 0.07 are given for mild and high strength steel, respectively.

The geometrical imperfections in the plate and the stiffeners is accounted for in the modelling of the ultimate moment capacity. In addition, a model uncertainty factor, X_R, is applied to the capacity, N(1.05, 0.1). The bias of 1.05 is based on a comparison between the single step method, which is applied here, and non-linear finite element analysis results by Törnqvist (2004).

6.2.6 Results of Structural Reliability Analysis

The wave climate as given in IACS Recommendation No. 34 (2000) is referred herein as a *Base Case*. The following increase of the extreme significant wave height is considered: 0.5, 1.0 and 2.0 m (see Sect. 2.6), which reflects the variation in the findings on climate change reported in the literature for the North Atlantic. The annual probability of failure for the tankers considered is presented in Figs. 6.4–6.8. The results are illustrated as a function of the steel deck cross sectional area, where modifications were implemented in terms of changes in the plate thickness and the stiffener size in a realistic proportional manner. The deck

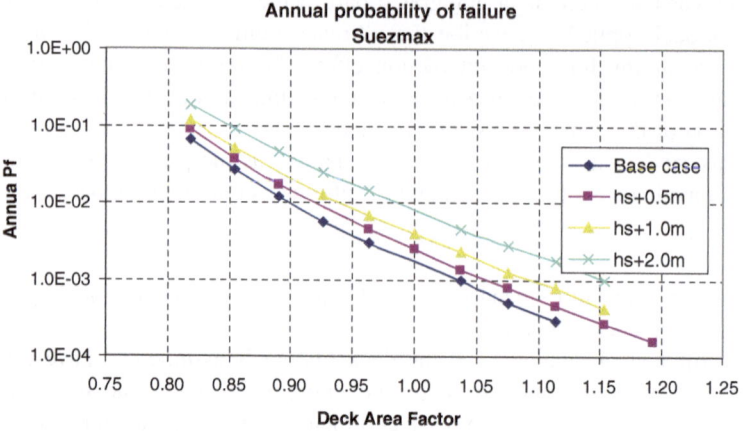

Fig. 6.4 Annual probability of failure, Suezmax

area factor equal 1 refers to the initial ship design without modification of significant wave height.

The calculated failure probabilities are nominal values, and should not be given a frequency of failure interpretation. The results are more useful on a comparative basis, and the absolute values should be interpreted with caution. In this context it should be noticed that the results presented here are for "net scantlings", which is part of the explanation of, what some would say, relatively high failure probabilities. Gross scantling, which is the net scantling plus the corrosion addition, would reduce the failure probabilities by approximately an order of magnitude. There may also be other reasons that the frequencies of hull girder failure in real

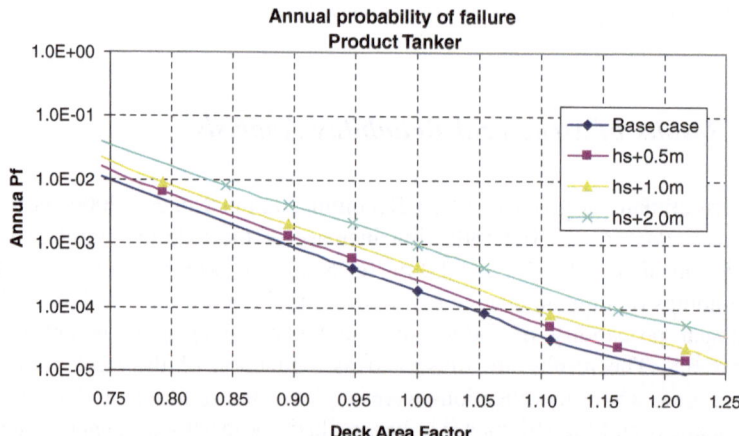

Fig. 6.5 Annual probability of failure, Product Tanker

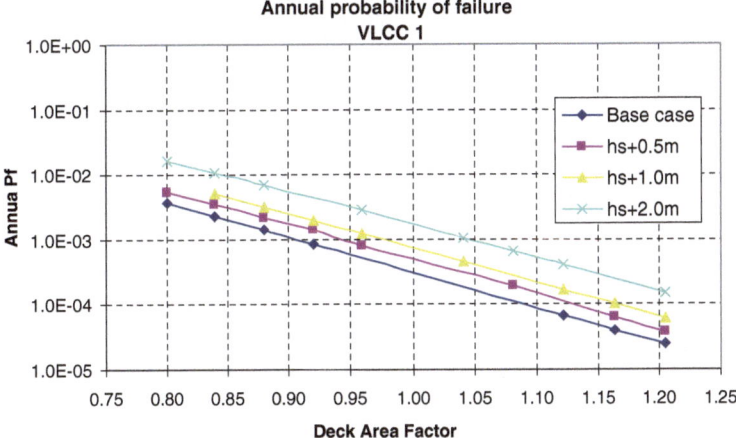

Fig. 6.6 Annual probability of failure, VLCC 1

life is lower than the calculated values reported here; e.g. weather routing is not accounted for, few ships operate in the North Atlantic throughout the lifetime, the yield strength is in many cases higher than the specification require.

The figures show the same overall trend for all the ships analysed. The probability of failure increases by around 50 % for each increase in H_s by 0.5 m. If H_s increases by 1.0 m, the deck area (steel weight of the deck in the midship region) needs to increase by some 5–8 % in order to maintain the reliability level. For increase of H_s by 2.0 m an increase the deck area of by 10–15 % will be needed. The results show that longer ships (VLCC 1, VLCC 2) seem to require the largest increase of the ship deck area.

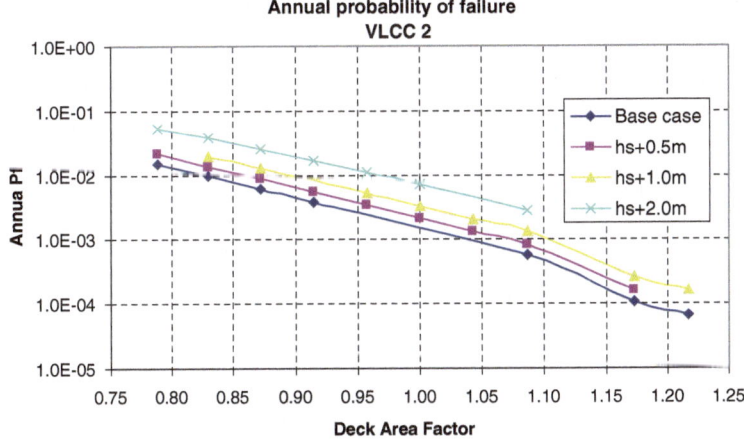

Fig. 6.7 Annual probability of failure, VLCC 2

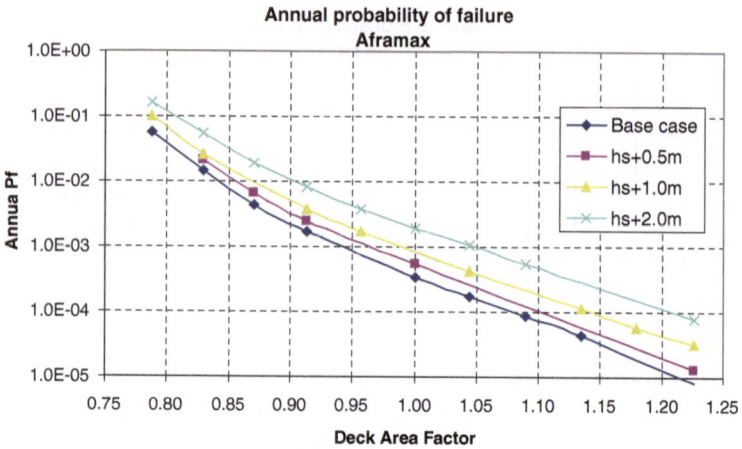

Fig. 6.8 Annual probability of failure, Aframax

Rough estimates indicate that reliability level is maintained in design if the characteristic wave bending moment, or the partial safety factor for the wave bending moment, is increased by ca. 8–10 % for an increase in H_s by 1.0 m. This increase is somewhat more than the increase in the deck area as discussed above.

References

Bach-Gansmo O, Lotsberg I (1989) Structural design criteria for a ship type floating production vessel PRADS 1989 conference proceedings

Bitner-Gregersen EM (1988) Joint environmental model. Annex A, det norske veritas report no. 87-31, Høvik. Madsen HO, Rooney P, Bitner-Gregersen E. Probabilistic calculation of design criteria for ultimate tether capacity of snorre TLP

Bitner-Gregersen EM, Haver S (1989) Joint long term description of environmental parameters for structural response calculations. In: Proceedings of the 2nd international workshop on wave hindcasting and forecasting, Vancouver, B. C

Bitner-Gregersen EM, Haver S (1991), Joint environmental model for reliability calculations. In: Proceedings of ISOPE'91 conference, vol 1, Edinburg, UK, pp 246–253

Bitner-Gregersen EM, Cramer E, Loseth R (1993) Uncertainties of load characteristics and fatigue damage of ship structures. In: Proceedings of OMAE 1993 conference in Glasgow, June 20–24 1993

Bitner-Gregersen EM, Cramer E (1994) Accuracy of the Global Wave Statistics data. In: Proceedings of ISOPE 1994 conference, Osaka, Japan, 10–15 April 1994

Bitner-Gregersen EM, Cramer E, Løseth R (1995a) Uncertainties of load characteristics and fatigue damage of ship structures. Mar Struct 8:97–117

Bitner-Gregersen EM, Cramer EH, Korbijn F (1995b) Environmental description for long-term load response of ship structures. In: Proceedings of ISOPE-95 conference, Hague, Netherlands, 11–16 June 1995

Bitner-Gregersen EM, Hovem L, Skjong R (2002) Implicit reliability of ship structures. In: Proceedings of OMAE 2002, Oslo, 23–28 June 2002

Bitner-Gregersen EM, Guedes Soares C (2007) Uncertainty of wave steepness prediction from global wave databases. In: Proceedings of MARSTRUCT conference, Glasgow, March 2007

Bitner-Gregersen EM, Skjong R (2011) Potential impact on climate change on tanker design. DNVRI Position Paper No. 8. http://www.dnv.com/resources/position_papers/impact_climate_change_tanker_design.asp

Bitner-Gregersen EM, Hørte TF, Skjong R (2011) Potential impact on climate change on tanker design. In: Proceedings of OMAE 2011 conference, Rotterdam, 19–23 June 2011

BMT (British Maritime Technology) (Hogben N, Da Cunha, LF, Oliver HN) (1986) Global wave statistics. Unwin Brothers Limited, London

Chen YRN, Thayamballi AK (1991) Consideration of global climatology and loading characteristics in fatigue damage assessment of ship structures. The marine structural inspection, maintenance and monitoring symposium. Virgina, March 1991

DNV (1992) Classification note 30.6: structural reliability analysis of marine structures. July 1992

DNV (2002) PROBAN theory, general purpose probabilistic analysis program, the author Tvedt L, version 4.4, Høvik, Norway

DNV (2010) RP-C205: environmental conditions and environmental loads, Høvik, Norway, April (2010). The 2007 revision available at Internet in 2010. The original was issued in 2000

DNV (2009) PULS 2.0, Panel ultimate limit state, Copyright Det Norske Veritas AS, P.O.Box 300, N-1322 Hovik, Norway

Gran S (1992) Short-term still-water load statistics. DNVR Report 92-2070

Guedes Soares C, Trovao MFS (1991) Influence of wave climate modelling on the long-term prediction of wave induced responses of ship structures. In: Price WG, Temarel P, Keane AJ (ed) Dynamics of vehicles and structures waves. Elsevier Science Publishers, Amsterdam

Hogben N, Lumb FE (1967) Ocean wave statistics. HMSO, UK

Hørte T, Skjong R, Friis-Hansen P, Teixeira AP, Viejo de Francisco F (2007a) Probabilistic methods applied to structural design and rule development. In: Proceedings of the RINA conference "Development of Classification & International Regulations". 24–25 Jan 2007, London

Hørte T, Wang G, White N (2007b) Calibration of the hull girder ultimate capacity criterion for double hull tankers. In: Proceedings 10th international symposium on practical design of ships and other floating structures. Houston, USA

IACS (2000) Standard wave data for direct wave load analysis. IACS recommendation No.34, Feb 2000. updated in 2001

IACS (2010) Common structural rules for double hull oil tankers with length 150 metres and above. Rules for classification of ships, part 8 chapter 1, July 2010

ISSC (2000a) Special Task Committee VI.1 Extreme Hull Girder Loading, vol 2

ISSC (2000b) Special Task Committee VI.2 Ultimate Hull Girder Strength, vol 2

ISSC (2009) Technical Committee I.1 "Environment". Bitner-Gregersen EM, Ellermann K, Ewans KC, Falzarano JM, Johnson MC, Nielsen Dam U, Nilva A, Queffeulou P, Smith TWP, Waseda, (Chairman of Committee I.1: E.M. Bitner-Gegersen),vol 1, pp 1–126

Kaminski LK (1997) Reliability analysis of FPSO's hull girder cross-sectional strength. In: Proceedings of OMAE 1997–vol II, safety and reliability ASME

Mathisen J Birknes J (2003) Statistics of short term response to waves first and second order modules for use with PROBAN. DNV report no. 2003-0051 Rev. No. 02, 07 Nov 2003

Mathisen J, Bitner-Gregersen EM (1990) Joint distribution for significant wave height and zero-crossing period. Appl Ocean Res 12(2):93–103

Skjong R, Bitner-Gregersen EM, Cramer E, Croker A, Hagen ø, Korneliussen G, Lacasse S, Lotsberg I, Nadim F, Ronold KO (1995) Guidelines for offshore structural reliability analysis—general. DNV report no. 95 – 2018

Steen E, (1996) Buckling of stiffened plates with application to longitudinal ship strength. Preliminary draft report, University of Oslo, Mechanics Department (PhD thesis published in 2001)

Turkstra CJ, (1970) Theory of structural safety. SM study no. 2 solid mechanics division, University of Waterloo, Waterloo

Törnqvist R (2004) Non-linear finite element analysis of hull girder collapse of a double hull tanker. DNV report no. 2004-0505, rev. 00, 13 May 2004

Vanem E, Bitner-Gregersen EM (2012) Stochastic modelling of long-term trends in the wave climate and its potential impact on ship structural loads. Appl Ocean Res 37(2012):235–248

WMO (World Meteorological Organization) (2001) Guide to marine meteorological services, 3rd edn. WMO-No. 471, Geneva

WMO (World Meteorological Organization) (2003) Manual on the global observing system, vol I and II. WMO-No. 544, Geneva

Chapter 7
Conclusions

The climate system is very complex and its mechanism is still not fully understood, however, observed and projected climate changes indicate that changes in met-ocean conditions can be expected. This will impact on ship and offshore structural design.

The results presented by IPCC (2007), 2012 are strongly dependent on an adopted scenario for emissions and concentration of CO_2 and are affected by various types of uncertainties which need further investigation.

This review has identified that there is an agreed increase in significant wave height from the middle of the twentieth century to the early twenty first century in the northern hemisphere winter in high latitudes in the north Atlantic and the north Pacific. There has been a decrease in more southerly latitudes of the northern hemisphere. The increase of the 99-percentile significant wave height has been observed to be up to 0.5 % per year (Young et al. 2011). However, if the record is extended back to late nineteenth century the picture changes, as studies show that storminess and wave heights in late nineteenth/early twentieth century were about the same as near the end of the twentieth century. Thus, it is unclear if the increase observed during the last 4–5 decades is caused by anthropogenic climate change or just manifestation of long-term natural variability.

The review has also identified that the future is likely to bring regional increases in the wind speeds and wave heights, more pronounced for the extremes than for the means. The increases of the 20-year return period of SWH or the highest storms in 20–30 years intervals are generally in the range 0.5–1.0 m in the North and Norwegian Seas, immediately west of the British Isles, off the northwest of Africa, around 30°N from the east coast of the United states to 50°W and in the Pacific between 25°N and 40°N and from the west coast of the United States to 170°W. However, increases up 18 % for the 99th percentile SHW have been reported for the southern North Sea by Grabemann and Weisse (2008). There are indications that the increase in extreme wave heights may reach more than 10 % above present day extremes in some areas. The projections are influenced by choice of climate model, emission scenario and downscaling method for waves. The uncertainty of the estimated increases is of the same order as the estimates.

E. M. Bitner-Gregersen et al., *Ship and Offshore Structure Design in Climate Change Perspective*, SpringerBriefs in Climate Studies, DOI: 10.1007/978-3-642-34138-0_7, The Author(s) 2013

Although the uncertainties in future projections of extreme wind speed and wave heights are less known than for surface temperature and precipitation they should not be ignored when impacts of climate change on design and operation of ship and offshore structures are considered.

In light of the findings summed up above the potential impacts of climate change on the tanker design has been investigated. The study indicates that observed and projected changes in wave climate will have large effects on tanker design practice. The presented examples show that in order to maintain the safety level the steel weight of the deck for net scantlings should be increased by 5–8 % if the extreme SWH increases by 1 m.

This calls for investigations of the necessary increase in partial safety factor(s) and/or revised specification of the characteristic wave bending.

An approach combining continuously new information about climate change and extreme weather events and relevant uncertainties in current design practice of ship and offshore structures is proposed. To be able to design for climate change statistical extreme value analysis, as currently used in the met-ocean community, has to be upgraded to take into account for the non-stationary character of current climate. This needs to include both anthropogenic climate change trends and natural variability cycles.

In the case of climate change leading to more extreme weather, rules for tankers would need revisions in order to maintain the structural reliability level. This could be done either by revising the IACS formula for the characteristic wave bending moment or by increasing the partial safety factor for the wave bending moment. Alternatively, one could consider introducing direct calculation of the characteristic wave bending moment and apply the environmental model corresponding to the climate change in this calculation. This has not been done in the present study.

Further studies are called for to describe and quantify potential implications of climate change on safe design and operations of ship and offshore structures as well as related economic consequences before firm conclusions are reached regarding possible updates of Classification Societies Rules and Offshore Standards.

The industry should continue to develop decision support systems which need to be associated with proper warning criteria to extreme weather events. However, for some phenomena, e.g. rogue waves, there is still a need for a better understanding of the actual phenomena, particularly met-ocean conditions when rogue waves occurred.

References

Grabemann I, Weisse R (2008) Climate change impact on extreme wave conditions in the North Sea: an ensemble study. Ocean Dyn 58:199–212

IPCC (2007) Climate change (2007). The physical science basis. Contribution of working group I to the fourth assessment report of the intergovernmental panel on climate change. Solomon S, Qin D, Manning M, Chen Z, Marquis M, Averyt KB, Tignor M and Miller HL (eds) Cambridge University Press, Cambridge, United Kingdom and New York, pp 996

IPCC (2012) Managing the risks of extreme events and disasters to advance climate change adaptation. A special report of working groups I and II of the intergovernmental panel on climate change. Field CB, Barros V, Stocker TF, Qin D, Dokken DJ, Ebi KL, Mastrandrea MD, Mach KJ, Plattner G-K, Allen SK, Tignor M and Midgley PM (eds) Cambridge University Press, Cambridge, United Kingdom and New York, pp 582

Young RI, Zieger S, Babanin AV (2011) Global trends in wind speed and wave height. Science 332:451–455 (22 April 2011)